Dedication

Dedicated to Richard Danne,
Bruce Blackburn, and the
Danne & Blackburn staff.

Richard Danne

My instinct is to look forward, rarely back, and my favorite day is tomorrow. Yet this NASA story... the Trek, refuses to go away.

All of us at Danne & Blackburn were fortunate to be involved in such an unusual and exciting program. And for me, it was a singular professional experience, loaded with struggle and success. Even as a kid on our Oklahoma farm, I recall lying in the grass at night, staring at the heavens, awed by the magnitude and mystery of what was – *out there!*

For Bruce and me, this was the chance to take our young firm to another level. With only five people, including the partners, it was "a reach" to tackle and deliver one of the nation's highest profile design programs. The federal arena was uncharted territory. And though it might look easy when viewing the completed Manual, unless you have redesigned a federal agency you can't imagine how complex and daunting it is. D&B eventually redesigned three more federal agencies, so I'm here to tell you.

Extremely low pay, countless hours, bureaucratic barriers, endless small contracts, and intense resistance from NASA's old-timers. This last fact is shocking to people who assume the agency was a wildly progressive, forward-thinking organization. Actually, NASA was a collection of eleven disparate centers, all competitive, all resisting Headquarters.

But the struggles tend to pale in the afterglow of the completed assignment. Bruce made a big contribution up front with the logo; Stephen Loges was the indispensable designer/colleague who helped push this unique program along; and I served as design director and designer throughout. It was a marathon trek, taking over ten years to complete Manual supplements and many special projects. I'm sure we would do it again.

This was important work, and our firm was given the opportunity to score an early victory for the National Endowment for the Arts, which fostered this federal initiative. We took the challenge very seriously and delivered. The successful NASA program then created momentum and propelled other Agencies to embark on their own redesigns. NEA program head Jerome Perlmutter called NASA "the flagship of our Federal Graphics Improvement Program."

In 1985, NASA was honored with a U.S. presidential Award for Design Excellence, in the very first year of such awards. In ceremonies held in Washington, and on behalf of Danne & Blackburn, I accepted it from President Reagan.

We never dwell on our program's being rescinded, yet there are constant reminders from the public. Every few weeks, another request arrives from a publisher, editor, or graduate design student who wants to feature and honor it. These inquiries come from around the globe, most often from Asia and Europe, and I respond each time by sending editorial and visual material.

So although there is that deep sense of loss, it is countered by the ongoing support of an international community. And it's thrilling to work with the next generation of designers, like the talented, dedicated Jesse Reed and Hamish Smyth. There is kinship, respect, and a shared vision of the past as future.

I remain an enthusiastic fan of NASA and of space exploration, and revel in each accomplishment of Hubble or *Voyager*. Plus there is satisfaction in knowing our spacecraft markings are flying – *out there* – in very deep space.

Onward!

Bruce Blackburn (left)
and Richard Danne at
their office at
One Dag Hammarskjöld
Plaza, New York City

Photo:
Alan Orling

Danne & Blackburn's
office at "One Dag,"
1975

Pictured:
Stephen Loges (right),
Ruth Donovan, and
Bruce Blackburn

Photo:
Bill Maris

There's an episode of *Star Trek: The Next Generation* titled "The Royale," first aired in 1989, in which the U.S.S. *Enterprise* encounters a mysterious field of wreckage orbiting a distant planet. The starship's crew beams a chunk of the debris aboard, and determines that it's a fragment of the *Charybdis*, a vessel launched from Earth in 2037 and lost thereafter. It's easy enough to identify, because the piece they take aboard displays two symbols: an American flag bearing 52 stars (dating it, they say, to the mid-21st century) and a logotype of four curvy letters reading N-A-S-A.

"The Royale" has been looping around the planet in reruns since it premiered. Richard D. James, the show's production designer, surely never imagined that his team was inadvertently setting up a big fat continuity error. But in 1992, just three years after the episode aired, the National Aeronautics and Space Administration dumped that sinuous logo and systematically began to strip it from buildings, documents, uniforms, and spacecraft, replacing it with an older insignia – nicknamed the Meatball – that the agency had retired nearly two decades before. It was a quixotically vigorous effort for something so symbolic, and the strange path that led there encounters the existential question graphic designers face: How important is what we do? And another question: How did the Meatball defeat the Worm?

FIRST STAGE: LAUNCH

The battle of the logos began before NASA, or for that matter *Star Trek*, existed. In 1957, the United States was making plans for its first space flight. As part of a deliberately apolitical program called the International Geophysical Year, a satellite would circle the Earth, gathering shareable scientific data during a stretch of peak solar activity. Rocketry, telemetry, and instrumentation had all advanced rapidly since the end of World War II, and it was perfectly reasonable to expect that research in the high atmosphere would move into orbit next. Project Vanguard had been announced and funded in 1955, and run by the U.S. Navy. The old National Advisory Committee for Aeronautics, known as NACA and founded in 1915, had been working on missile technology as well. Its first launch, initially scheduled for September 1957, was delayed till spring.

On October 4, 1957, the Soviet Union announced the launch of *Sputnik*, its own satellite. It was a polished sphere with four antennae trailing off it, and it functioned as a beacon but little more; its only signal was a steady beep. If you had a shortwave radio, you could hear it passing overhead. The Soviets were proud of their achievement, and although Nikita Khrushchev's government issued some propagandistic expressions of pride, they were not outlandish. This was a nice little scientific move, to be followed by bigger and better efforts, same as the American one. *Sputnik* beeped along for just 21 days, until its batteries died. Ten weeks after that, it burned to bits as it fell back into the atmosphere. The American intelligence agencies had known it was going to be launched, and no panic button had been pressed.

The rest of America, though, went bananas. The Communists! Those plodders who were okay at building tractors and tanks but nothing high-tech – they were conquering the skies, the moon, the heavens themselves. Everyone was talking in military terms, not scientific ones. (Tom Wolfe, in his definitive *The Right Stuff*, notes how many public figures used the phrase "capturing the high ground.") RED BABY MOON, read a giant headline in the New York *Daily News*. The usually more demure *Herald Tribune*, across all eight columns of its front page, went with a banner reading U.S. VIEWS SATELLITE AS RUSSIAN VICTORY. Slightly cooler heads saw the point of its launch vehicle: The Soviet rocket that put a satellite 130 miles into space had been developed to take a hydrogen bomb across the Pacific. And, after all, researchers in the United States had been quietly thinking about how to build a spy satellite. Surely we weren't the only ones.

A month later, the Soviets outdid themselves, sending up *Sputnik 2*, a half-ton capsule containing a dog named Laika whose (final) life signs were monitored by telemetry. When the United States tried to launch its own *Vanguard TV3* satellite a few weeks later, in December, the rocket blew up on the launch pad. The Americans were put in the unfamiliar position of looking hapless and inept.

Congress, antagonized and panicky, scrambled to offer money and support, and within just a few weeks passed the National Aeronautics and Space Act of 1958, forming a new civilian organization devoted to space flight. The new National Aeronautics and Space Agency incorporated NACA's operations, and finally got a satellite up the following May, a little thing called *Explorer 1*. The space race was on.

Few people in those early days of NASA were thinking about graphic design. The agency was run by a mix of military folks, flyboys, and bureaucrats, and not a lot of them were concerned with the field that was more often called "commercial art." But NASA did need something to put on its letterhead, and its administrator, T. Keith Glennan, asked a man named James Modarelli to work up both a formal seal (the kind that could be embossed or gold-stamped on a certificate) and an everyday insignia that could serve as a logo for the agency. NACA's old symbol, a shield with feather-edged wings, was ill-suited to represent space flight.

Modarelli, a technical illustrator at NACA, had risen to run the research-reports division at the agency's Lewis Flight Propulsion Laboratories, and now did the same at NASA. He was well-liked; to this day, he's fondly recalled by his old colleagues. He had thoughtful ideas about what might make a good logotype, including a curved model of a delta-shaped wing that he'd seen at a meeting a few months earlier. There were multiple such airfoils being tested in the labs, and an image of one of them made its way into a sketch Modarelli offered. A further call for submissions from the rest of the agency brought in 350 sketches and designs. When judging time rolled around, Glennan (according to the comprehensive research done by Joseph Chambers, a longtime employee at NASA's Langley Center, now retired) listened to half a day's worth of arguments, then exasperatedly declared that he was just going to choose one he liked. It was Modarelli's own.

The seal is rarely seen outside NASA, but it's still in use on certain official paperwork and ceremonial material, and appears, in black-and-white, on page 45 of this book. It shows a red airfoil, straining upwards, straddling and casting a shadow upon a yellow planet, set against a blue-black starry sky. The whole thing was surrounded by a ring of type reading NATIONAL AERONAUTICS AND SPACE ADMINISTRATION U.S.A.

In fact, streamlined parabolic boomerang shapes like this were propagating throughout American design just then, hinting at shock waves and supersonic travel. Speedo, the swimsuit company, had been using one for years, to connote hydrodynamic efficiency. In 1955, Chrysler had adopted a pair of boomerangs, one laid over the other, as part of its "Forward Look" campaign. The Hamilton Watch Company's new line, the very first wristwatches to run on batteries, included boomerang-shaped models called the Ventura and the Pacer. Tables and chairs with slim steel "hairpin" legs showed off similar forms. So did Formica countertops, bowling-alley signs, kitchen drawer pulls, even ashtrays. Literally and figuratively, sonic booms were in the air.

The design for the seal had to go through a long chain of approval. Most of the review came from an ancient-sounding agency called the Heraldic Branch of the Army Institute of the Quartermaster General, followed by the government's Commission on Fine Arts and finally (implausibly) by President Eisenhower himself. The first drawing submitted to the approval process had to be yanked back and redone because the red delta wing had been drawn upside-down.

Modarelli's seal probably wouldn't be okayed today. It's not much of a design. None of the elements hang together; the lettering is badly handled and a little askew. Over the course of its history, the type changed from a sans serif to a serif face, which made it worse rather than better. The whole thing looks like a piece of clip art. Nonetheless, the Army and the Commission and the president all signed off on it in 1959. When it appeared on a commendation given to Alan Shepard after his first flight, *Time* magazine offered that the award "looked as if it might have come out of a Cracker Jack box. The Distinguished Service Medal of the National Aeronautics and Space Administration is the most unimaginative decoration turned out by the U.S. government so far – and the competition for that title is stiff." There are NASA old-timers who are still upset about that *Time* story, but the design hasn't improved any with age.

The seal was also too complicated and fine-grained for widespread use, and Glennan's design brief had called for a simplified insignia as well. Unlike the seal, it would serve everywhere, from stationery to uniform patches to the exteriors of hangars and spacecraft. It also wouldn't have to clear the Army's Heraldic Branch, just the fine-arts group.

You know what it looks like. The logotype that NASA adopted shortly thereafter – and uses today – bore a relationship to the seal, but was far better and more coherent. The field was a uniform disc of blue (corresponding to Pantone 286, if you happen to be specifying it), with the acronym and a scattering of stars knocked out in white. The letters were orbited by a white satellite on a canted, elliptical path. The red delta wing from the seal (Pantone 185) again jockeys atop it all, bigger this time. The four letters, with big chiseled Latin serifs, were chunky and muscular, befitting the agency's toughness. It was, unlike the seal, a coherent design, if a busy one. It was accepted by the arts commission, though reluctantly, with a note about its "limited sculptural possibilities."

It also had significant practical flaws, which seems surprising given NASA's technological bent. At a distance, the white letters of the logotype grew hard to read against the star field. When it was printed small, it turned into a blob. In black-and-white documents – which were commonplace – it looked like "a thumbprint," one designer recalls. Government-issue printing in those days was pretty bad, and when the mix of inks was slightly off, the background blue turned to mud. As the arts commission had feared, it didn't look very good in 3-D either.

But it resonated within the military and engineering culture of NASA, and perhaps beyond. It offered speed and contrails and the skies themselves; it was not an abstraction but a literal depiction of the agency's mission. It was a boy adventurer's idea of space travel, and quite a few of those boy adventurers had grown up to become the pilots and scientists constituting this new agency and pushing, hard, into space. It spoke to them, and the logo soon became a ubiquitous presence at the agency's centers, making its way onto aircraft wings, rocket-booster stages, and flight suits. It was painted on buildings at giant size.

Over the next decade, NASA had, it's fair to say, more on its mind than a mildly problematic logo. In 1961, John Kennedy laid out a plan to put a man on the Moon and bring him back safely by 1970. It was a shrewd way of funding vast developments in rocket technology – which was to say, ballistic-missile technology – through a program that even some pacifists believed in. It also leveled the playing field with the Soviet Union, because although the U.S.S.R. already had rockets that could put a man in orbit and were somewhat more advanced than the American ones, neither side had a multistage booster that could get to the moon. If both sides were starting that program from scratch, American industry could probably outpace the Soviet collective.

After JFK was killed a year and a half later, Cape Canaveral became Cape Kennedy, and the push to the Moon became, in part, a tribute to him. Even when the astronauts Roger Chaffee, Edward White, and Gus Grissom died in a horrible launch-pad fire in January 1967 – after which the Moon-bound command module had to be redesigned essentially from scratch – the program never stopped surging forward, in an effort (and at a financial cost) that seems impossible today. On July 20, 1969, Neil Armstrong and Buzz Aldrin stepped out onto the Sea of Tranquility, slightly north of the lunar equator, and left on the dusty surface a planted American flag and a plaque reading WE CAME IN PEACE FOR ALL MANKIND. The NASA insignia appeared on their spacesuits, front and back. Though he probably never expected it, James Modarelli had become the first graphic designer to leave his mark on another world.

SECOND STAGE: APOGEE

By 1972, the mood surrounding NASA – and within it – was beginning to shift. Though there was a lot of good science going on at the agency, its blazing-eyed sense of mission was beginning to wane. The semi-autonomous component organizations that made up NASA – the Jet Propulsion Laboratory in Pasadena, the Johnson Space Center in Houston, the Kennedy Space Center in south Florida, and eight others – were still, even after more than a decade, chafing at the overarching decisions made at NASA headquarters in Washington. The *Viking* and *Voyager* missions to Mars and the outer planets would not arrive for another couple of years, so there were no great pictures coming back from far-off worlds. The first big economic slowdown since World War II, twinned with the newly empowered counterculture, had left a lot of people questioning NASA's budget, saying things like "maybe we'd better solve some problems back on Earth first."

Without a hard goal – the Moon, by order of a martyred president – in sight, the agency was for the first time being asked to justify itself. The final three of the ten Apollo missions to the moon were canceled to save money. Many of the remaining Apollo parts were reassembled into, and used to supply, a prototype space station called Skylab. Though it produced valuable research, and kept three crews in space for, collectively, nearly six months, it failed to catch the public eye in a dramatic way. Its third crew got fed up with Mission Control's tight scheduling and staged a revolt in mid-mission, effectively going on strike until the astronauts were granted more free time and less micromanagement. The most attention Skylab ever got from the public was probably when it fell out of orbit and crashed to Earth in July 1979.

The graphic-design business was, in these years, reaching a sort of peak. The high modernism that had dominated the visual arts – architecture, graphics, fine art – was nearing its zenith of ubiquity and reach. Virtually all big corporations represented themselves with severe geometric forms, and most had finally come to appreciate that good graphics could have extraordinary power. Designers were supplying clients with not just letterheads and logotypes but a full set of treatments collectively known as a "graphic identity." A few years earlier, the largest communications company ever, AT&T, had asked Saul Bass to revamp its Bell System logo, to which it had made only incremental changes for a century, and Bass redid everything down to the linemen's uniforms. Coca-Cola had begun to play down its signature Spencerian script and Victorian hobble-skirt bottle in favor of the blocky word "Coke" atop a twisty abstracted ribbon. In a weird turn of history, Bauhaus-style modernism, created by a bunch of Socialists in 1920s Europe, had been repurposed as the dominant look of American corporate power.

The federal government was, as you might expect, a sluggish latecomer to that change. Its agencies' graphic identities were inconsistent, some formal, some nonexistent. Even its physical paperwork was behind the times: Owing to a decision made in the 1920s, most government correspondence took place on a nonstandard paper size, eight inches by ten and a half, instead of the eight-and-a-half-by-eleven sheet that the rest of the country had settled upon. The letters people received from the government often looked like they, too, were from 1925.

The contemporary-graphic-design banner was borne in by a woman named Nancy Hanks. She was a blue-blood, descended from Nancy Hanks Lincoln, mother of the sixteenth president, and in 1969 was appointed by Richard Nixon to run the National Endowment for the Arts. The NEA was a new organization – she was only its second chairperson – and it was tiny, with a $16 million budget. Hanks, though, had big aims, and fine political skills. She was a moderate Rockefeller Republican, one who could talk persuasively and sanely to Democrats. When, in 1970, the agency was threatened with defunding, she is said to have personally coaxed 100 senators and congresspeople to have switched their votes (And by the time her term ended eight years later, the NEA budget was $100 million – which is, in inflation-adjusted dollars, more than the agency gets today.)

Hanks grasped that a graphic-identity program had the potential to do for government agencies what it had done for corporations: convey modernity rather than stodginess, and forward-looking and efficient practices rather than hidebound ones. As Lani Lattin, her deputy, recalled after her death, "We said that if government publications – from tax returns to road signs – were attractive and easy to read, people would read them and respect them. We said that the morale and the productivity of our public servants would rise dramatically if they could work in a decent environment, instead of sitting at tacky desks staring at hideous green walls." They had the backing of the presidential aide Leonard Garment, and by extension the Nixon White House.

In 1972, Hanks launched the Federal Design Improvement Program, a series of grants allowing government departments to hire architects and designers who could supply those better-looking desks and reports. The agencies themselves, not the NEA, handled the proposals and decisions, though the graphics program was overseen by an NEA executive named Jerome Perlmutter. Early on, the Department of Labor received (among other upgrades) a good-looking logotype of meshing L-shapes that evoked the Stars and Stripes. NASA was likewise early on the list, and in 1974, the Request for Proposals went out.

Danne & Blackburn's submission was, one would think, a long shot. The firm was very small, without the cachet of Vignelli Associates or Paul Rand or the Eames Office or any of the other prominent design shops of the era. It was barely a year old, with three employees. But its two partners, Richard Danne and Bruce Blackburn, knew what they were doing: Danne had been doing work for General Dynamics and (with his colleague Phil Gips) a lot of Paramount movie posters, including a superb one for *Rosemary's Baby* in 1968. Blackburn had a foundation in corporate-identity work, from time spent at the prominent firm Chermayeff & Geismar, and had just won the competition to create a symbol for the United States Bicentennial celebration. Perlmutter had seen and admired that logo, and he had called Danne & Blackburn to say that the RFP would be going out shortly, and that they were well-positioned to apply. The proposal they sent in was badly typed and full of Wite-Out. It also (amazingly enough) completely lacked illustrations. But it was good enough to get them a chance to show what they could do. After about a month, they heard from NASA that they'd got the job.

The two designers pretty quickly realized that the problem was not just one of type sizes and white space but of administration. A good graphics program could unify the eleven NASA centers, at least visually, and at the same time systematize all the documents, signs, and other materials they produced, which at the time varied from crude to nondescript. "We really looked at how we were going to pull all the centers under the banner and link them very strongly with headquarters," Danne says today. "This was not a popular solution out in the field, but that was really at the core of the thing. And then we started working with pictorial symbols – not like the Meatball but with more going on like that – and what we realized was that with so much chaos, we needed something terribly simple to anchor our program."

That something came out of Blackburn's pen – and it was a pen, this being a pre-Macintosh age. He drew dozens of variants. (Danne notes, admiringly, that "Bruce worked on this heavily – he was really facile, really fast, and could just push 'em out.") The drawings grew simpler, because, he says, "we were backing in, in a defense stance, and we said, well, obviously we have to make it more modern." They settled on a monochrome logo with a single-width stroke, constructed of just three compound-curve lines. It was bold, like NASA itself; it was technological in its swoop. The two capital *A*s lacked crossbars, suggesting rocket nose cones or the shock waves that those nose cones produced. With a couple of lines of black Helvetica type beneath or alongside that sturdy logo, a system for neatly differentiating the agency's discrete centers snapped into place.

Though the logotype could be produced in black or knocked out of a color in white, it was most often specified in a solid orangey red corresponding to Pantone 179, the same color as the delta wing from the earlier insignia. (They gave it a name, NASA Red, that the agency still uses.) The form worked at large and small sizes, and would hold up even on the worst printing press or photocopier. The color was a slightly unusual choice, but one that Danne and Blackburn believed in. Airlines and military agencies tended to go with blue logos, suggesting sky and sea and coolness; they instead chose to convey action and heat.

Danne and Blackburn brought their work to NASA together in October 1974. A decent-sized group attended the presentation, but really the two opinions in the room that mattered were those of Dr. James Fletcher, the NASA administrator, and George Low, his deputy. Fletcher was the leader who was reshaping the agency's mission to address some of those earthbound problems that taxpayers were grumbling about; he had also been the man who had successfully persuaded the Nixon administration and Congress to back the Space Shuttle program, and got the *Voyager* and *Viking* probes under way as well. He was an abstemious Mormon who was open enough to work within a culture of chain-smoking engineers and hard-drinking pilots, and he was known as a cautious, thoughtful scientist.

At the meeting, more than two dozen display boards (reproduced at the end of this volume) showed the logo in action, in many settings. Danne and Blackburn finished their talk, whereupon Fletcher and Low began to offer feedback. The conversation very quickly ended up taking a turn that could have been written by Aaron Sorkin for an episode of *The West Wing*. As Danne remembers it:

FLETCHER:
"I'm simply not comfortable with those letters. Something is missing."

LOW:
"Well, yes, the cross stroke is gone from the letter A."

FLETCHER:
"Yes, and that bothers me."

LOW:
"Why?"

FLETCHER:
[*Long pause*] "I just don't feel we are getting our money's worth!"

And then, a few minutes later:

FLETCHER:
"And this color, red, it doesn't make much sense to me."

LOW:
"What would be better?"

FLETCHER:
"Blue makes more sense ... Space is blue."

LOW:
"No, Dr. Fletcher, space is *black*!"

NASA seal,
1958-present

Design:
James Modarelli

NASA insignia,
known as
the "Meatball,"
1958-1975,
1992-present

Design:
James Modarelli

NASA logotype,
known as
the "Worm,"
1975-1992

Design:
Danne & Blackburn

Mind you, this was not a rancorous argument. Fletcher probably intended to convey that the minimal design overall, and not specifically the missing crossbars, struck him as underdeveloped rather than merely clean. However his hesitance was meant, though, the conversation cast the final decision in an uncertain light. On the way out, Blackburn and Danne put their chances at 50-50. Back at the office, Blackburn even tried redrawing the logo with the crossbars restored to each *A*, though it didn't work at all. Only later did they get the call, which came in the form of a letter in the U.S. Mail: In NASA's favored language, it was a go.

The next weeks saw their little shop producing the original version of the manual you see here. Though much of its space explains how to prepare documents for publication, every application was worked out, from brochures to automobile door decals. Because it was so straightforward, the new logotype could be painted on a big building without losing impact. Its horizontal shape, unlike the Meatball's, lent itself to placement on airplane fuselages and the like – but not on rockets, because standing it on end was prohibited.

Many corporations and agencies had graphics manuals like this, but this is almost surely the only one that specifies the orientation of a logo when it's placed on a spacecraft. (I suppose it's possible that the Soviets had a graphics manual of their own. Probably not.) Danne remembers that, when the NASA lettering was stenciled on the Space Shuttle a couple of years later, it was required to be smaller than the words UNITED STATES, and that there were significant limitations on where it could be placed, owing to the various heat-resistant properties of the vehicle's tiled skin. "We weren't going to fight over that one and cause a crash," he recalls, with a chuckle.

Every image on the book's 90 pages was hand-drawn with India ink and French curves and T-squares, using the laborious, wearying techniques that were then taught in many high schools. Today, half a lifetime later, they are all but lost arts, and although some older designers say they miss them, nobody's going back anytime soon. One inkblot and you had to start the whole drawing over.

By the time the new program was rolled out, a year later, Fletcher – at least publicly – had embraced the new graphics. "I think the new logotype is pleasing to the eye and gives a feeling of unity, technological precision, thrust and orientation toward the future," he wrote in a letter, seen on page 25 of this book, that introduced Blackburn and Danne's work. "I think we were fortunate in recognizing that our graphics could stand improvement; I am confident that the program we now have underway will be second to none." And, perhaps foreseeing what was to come, he added this: "In order to succeed, a program which departs from the accustomed must have the full support of every NASA employee. Top-level management must take the lead, our experts in the field of graphic design must follow, and all of us must see that the specifics are diligently monitored to insure that standards of excellence are maintained." Fletcher knew his agency well, and it sounds like he had picked up on the blowback that was beginning to form.

Engineers, to make a gross generalization, are rationalists. The guy-with-a-pocket-protector-and-a-terrible-wardrobe cliché is, while dated, also telling: Comparatively few are aesthetes. (Though the best engineers frequently are.) They do things for good reason, not for good looks. Ask an architect sometime about the engineers who have compromised his or her elegant designs; then ask those engineers what they think of the absurd drawings they were asked to turn into a real building. Chances are, everyone will have a lot to say.

Danne and Blackburn had justified their case well to Fletcher and Low, but the men who did most of NASA's work at the various NASA centers had not seen their presentation or been told about the program. Instead of briefing them, NASA in 1975 sent their top executives a surprise: new stationery with the new logo. Perhaps because they'd been left out of the process, they were irritated, and their disdain found a focus right away, because some of them flat-out hated the new logo. The new symbol, after all, lacked the blue of the sky, the red-white-and-blue of the United States of America. To the hardest-headed literalists, it represented money foolishly spent on décor instead of on flight. They – like quite a few Americans – were suspicious of the spare aesthetics of high modernism, and although they didn't necessarily articulate it, they preferred more literal symbols to abstracted ones. To a certain populist view, design like this felt corporate and soulless rather than romantic and animate. There was also a noticeable generational divide: Younger employees liked the new logo, and older ones, by and large, did not. The administrators of the individual NASA centers felt that they were, once again, being forced in some stupid new direction by Washington. To them, it was yet another sign that, having achieved the Moon, the whole outfit was going blah.

From the beginning, NASA folks have had a knack for nicknames. Each class of astronauts, for example, gets one, from the Original Seven (1959) to the Maggots (1984) to the Chumps (2009). Very quickly, the new logo got an unflattering moniker of its own: the Worm. It was not affectionate, and the word was often pronounced with a disdainful sneer. Modarelli's old insignia, which had up till then been called "the Insignia," promptly got a nickname of its own. Credit for that goes to Frank (Red) Rowsome, the head of technical publications at NASA headquarters in Washington: He's the one who started calling it "the Meatball." And really, which would you prefer to represent your mission: something that was hearty, Mediterranean, and delicious, or something slimy, subterranean, and inedible? The Meatball sounded friendly; the Worm was sinister.

Fletcher had a backlash on his hands, and eventually sent Blackburn and Danne themselves on a tour of NASA campuses, in order to re-sell their work to the local administrators and quell the outrage. Danne recalls the visit to Houston, where Chris Kraft – the flight director of Mission Control during many of the space program's triumphs – was running the Johnson Space Center. Kraft had been intensely annoyed about the redesign, and sat through Danne and Blackburn's presentation skeptically. But at the end, he came around, saying, "Why wasn't it handled this way from the start? I don't have to like it, but I can see it's a real program, and I'm okay with it now." The designers' reasoned approach hadn't quite won everyone over, but it was enough to make the program stick. As did the nicknames.

Outside NASA, at least, the program was considered a big success. When the Reagan administration gave out awards for the best of these federal redesign programs several years later, NASA's was the first winner. Danne & Blackburn got to work with three more federal agencies. The Department of Transportation, its subsidiary the Federal Aviation Administration, and the Army Corps of Engineers all use D&B logos and graphics to this day. Saturn, the new General Motors division, launched in 1985 with a logo that mimicked the Worm so closely that NASA's lawyers got involved.

THIRD STAGE: RE-ENTRY

If the mood at NASA had seemed to be growing uneasy in the early 1970s, it was downright gloomy by 1992. The Space Shuttle program, though embraced by the public, had in 1986 seen its first fatal accident in space, the *Challenger* explosion. The agency had spent more than two years grounded, going through a deep self-examination and mourning period. The post-accident inquiry revealed that the agency's engineering had, at times, been stymied by pernicious and inbuilt bureaucracy. And as promising as the Space Shuttle had seemed, it had flown much less often than expected, at far greater cost. By the 1990s, its design was also two decades old and continuing to age. Flaws in the Hubble Space Telescope that it had launched had embarrassed the agency. Spaceflight in low Earth orbit, though highly useful, was simply not as sexy as tooling around on the Moon.

Dan Goldin took over this unhappy organization in 1992. Most of his goals involved slimming down the agency, which had been perceived as scattershot, slow-moving, spendthrift, and drifty, trying to carry off too many projects of hazy intent. He also aimed to shift many of NASA's traditional functions to the private sector. The jury is still out on his administration's long-term effects, but most NASA-watchers will agree that he was tremendously good at carrying out those goals.

American graphic design, it's not much of an exaggeration to say, was going through a sharp backlash in these years. If the early '70s had been the zenith of modern design, the early '90s were the peak of postmodernism, wherein pop culture was referenced, cut up, and reshaped, sometimes to the point of absurd incoherence. Even when things didn't go that far, the crispness of high modernism was giving way to busier, more ornate designs that made reference to older forms. (The explosion in the number of typefaces available to designers, owing to the desktop-publishing revolution, probably contributed to this.) Often these references were deployed with irony, but sometimes they were pretty straightforward. For its reworked graphic identity, Apple Computer commissioned a version of Garamond, a typeface whose origins date to the fifteenth century. Coca-Cola re-embraced its script and full name. Twirls of serif type, often awkwardly

NASA 83-HC-652

NASA photo caption:

15 October 1983

83-HC-652
83-H-765
S-83-40555

JOHNSON SPACE CENTER, HOUSTON, TX

STS-11 CREW
These five astronauts ar in training for NASA's 41B (STS-11) mission, scheduled early next year. On the front row are Vance D. Brand, commander; and Robert L. Gibson, pilot. Mission specialists (back row, left to right) are Robert L. Stewart, Dr. Ronald E. McNair and Bruce McCandless, II. Stewart and McCandless are wearing extravehicular mobility units (EMU) spacesuits. The STS program's second extravehicular activity (EVA) is to be performed on this flight, largely as a rehearsal for a scheduled repair visit to the Solar Maximum satellite, on a later mission. The manned maneuvering unit (MMU) will make its space debu on 41B.

Photo:
NASA

deployed, seemed to be replacing Helvetica on every package in the supermarket. Massimo Vignelli's bright orange gave way to Michael Graves's soft green. People seemed willing to accept almost anything except rectilinear lines and smooth color fields. Graphics like the Worm were often, and shortsightedly, put in the same class with elephant bell bottoms and seven-inch-wide lapels: embarrassing hangovers from the Decade That Taste Forgot.

One day, just after he took over, Goldin was flying into the Langley Research Center. Someone onboard the plane – possibly Goldin himself; there are conflicting versions of the scene – remarked on a piece of signage, wondering "why in the world we have that awful logo." Once they landed, he was met by Paul Holloway, who ran Langley, and Holloway offered a similar gripe. (He was a hardcore Meatball guy, going so far as to demand business cards with the old logo in place of the Worm.) As Goldin recently recalled to the New York *Times*:

> "He said to me, 'Do you see that?' I said, 'Yeah, that's the NASA logo.' He said, 'That's the Worm. ... If you want to really excite NASA employees about changes coming, why don't you tell them we're going to deworm NASA and bring back the Meatball?' "

Goldin turned to an official and asked, "Can I do that?" "Of course," came the response. And in Goldin's next speech at Langley, he announced that the Meatball was coming back, to enormous cheers. (His actual words were "the affectionate Meatball will replace the slick NASA logo, and slowly it will die into the horizon and never be seen again," which tells you something about the prevailing attitude.) The romantics had finally shoved back the aesthetes, as they'd wanted to for nearly two decades. The Worm had been exterminated. Danne & Blackburn had disbanded amicably in 1985; Danne was the one who got the call from his NASA contact, sorrowfully telling him that their standards program had been retired. (Nevertheless, many de-Wormed elements of it linger in NASA graphics to this day.) Modarelli, the next year, received an official commendation.

Over the next few years, the Worm was stripped from printed material, signage, and everything else. Stories abound at NASA about the frantic scrubbing of the Worm from any building or office that Goldin was about to visit, and about his irritation whenever he'd run across one that hadn't been removed. It was not a small effort, and in fact, if the Worm had hung on a few years more, it probably wouldn't have been eradicated at all. A logo change costs money to execute – new stationery, new signage, lots of labor – and Goldin himself soon set in place a philosophy at the agency, known as "better-faster-cheaper," that governs it to this day. Given NASA's latter-day constraints, the "cheaper" would probably have involved standing pat.

Which, so far, it has done since 1992. Yet the Worm still pokes its head out now and again. It's still up there in orbit, where it's painted on many satellites that have been zinging around the Earth for decades, and nobody's going up there to repaint them. It's chiseled in granite at the entrance to the NASA headquarters building in Washington, and has not been jackhammered off. The Space Shuttle orbiters were repainted in the late '90s, but the *Enterprise* – the proto-Shuttle used for landing tests, now in retirement at the Intrepid Sea, Air & Space Museum in New York – still displays its Worm.

Besides, the Meatball is not a failure. It is – as was pointed out to me by Hamish Smyth, who worked with his colleague Jesse Reed to produce the beautifully made volume you're reading now – a first-rate uniform patch. Even if you're a believer in the virtues of the Worm, the Meatball has a place in the NASA graphics universe. Government printing is vastly better than it used to be, and the blue now reproduces pretty well most of the time. When it's rendered well, the Meatball tells a story, and provides continuity. But it's hard to argue, from a design standpoint, that it's really better than the Worm. The Worm is forever modern; the Meatball, despite its charms, will never be. For an agency whose mission is so thoroughly directed at the future, a retro emblem is an eccentric choice at best.

In fact, the agency has begun to use a simplified, thinned-out version, with a white background that makes it much more versatile and reproducible, in certain applications like aircraft tails. (NASA calls this the Swoosh; I think of it as the Vegetarian Meatball.) Today's edition of the NASA graphics manual devotes quite a few pages to handling the Meatball itself on light and dark backgrounds and in single-color versions, which can get tricky. It specifies that it should never be printed smaller than five-eighths of an inch in diameter, lest it blot itself out. A small box also shows the Worm, with a stern note: "Use of the retired NASA logo requires permission from the Assistant Administrator for Public Affairs or designee. The retired NASA logo and insignia can never be used together." That may be, in part, because there have been attempts to graft the Worm's letterforms onto the Meatball's orb. The Wormball, some people call this. It's terrible.

Could there be one more reversal ahead? Given the cost it's unlikely, but it's also not impossible to imagine. To the post-postmodern eye, the Worm looks a lot cooler and less dated than it did when Dan Goldin cast his eye upon it that day at Langley; the Meatball, by contrast, looks increasingly like kitsch, and the starry speckles in the background mean that it reproduces badly on a computer or phone screen, where so many graphics live these days.

In 2015, when Matt Damon found himself stranded on Mars in the blockbuster film *The Martian*, his spacesuit bore a logo for a fictional spacecraft, the *Ares III*, on its shoulder. Its letterforms were a deep orangey-red, made of curvilinear single-width strokes, with the crossbeam of the A absent. It was unmistakably a neo-Worm, and it looked not retro but once again futuristic. If we're headed to Mars in the not-too-distant future, might the agency see a new (or new-old) logo as a way to rebrand itself once more? Could a spacecraft painted in NASA Red land on the Red Planet? If so, that *Star Trek* episode, the one set in 2037, will turn out to be right after all. The 52-star flag is going to be the tricky part.

--

New York City
November 2015

Further reading

Chambers, Joseph R., and Mark A. Chambers. *Emblems of Exploration: Logos of the NACA and NASA*. Washington: National Air and Space Administration Office of Communications, 2015. www.nasa.gov/sites/default/files/atoms/files/emblems_of_exploration.pdf

Chang, Kenneth. "$79 for an Out-of-Date Book About a Modern NASA Logo." *The New York Times*. September 1, 2015.

Danne, Richard. *Dust Bowl to Gotham*. San Francisco: Blurb, 2015.

Hine, Thomas. *The Great Funk: Falling Apart and Coming Together (On a Shag Rug) in the Seventies*. New York: Sarah Crichton Books, 2007.

Hine, Thomas. *Populuxe*. New York: Alfred A. Knopf, 1986.

Memory Alpha. "The Royale" (episode). Accessed November 21, 2015. www.memory-alpha.wikia.com/wiki/The_Royale_(episode)

National Endowment for the Arts. "Setting the Standard: The NEA Initiates the Federal Design Improvement Program." Accessed November 21, 2015. www.arts.gov/article/setting-standard-nea-initiates-federal-design-improvement-program#sthash.LMcFjjyY.dpuf

Straight, Michael. Nancy Hanks: *An Intimate Portrait: The Creation of a National Commitment to the Arts*. Durham and London: Duke University Press, 1988.

Tufte, Edward. *Beautiful Evidence*. Cheshire, Connecticut: Graphics Press, 2006.

Wolfe, Tom. *The Right Stuff*. New York: Farrar, Straus & Giroux, 1979.

Following pages

National Aeronautics and
Space Administration
Graphics Standards Manual

Designed by Danne & Blackburn
New York, 1975

National Aeronautics and
Space Administration

Graphics Standards Manual

NHB 1430.2
January, 1976

NASA

Introduction

Index

The following pages of the NASA Graphics Standards Manual, listed in sequential order, are officially in effect as of January 1, 1976

Index

The following pages of the NASA Graphics Standards Manual, listed in sequential order, are officially in effect as of January 1, 1976.

Dear Colleagues:

A driving force and the use of innovative techniques and ideas have brought NASA the image of a get-it-done agency, and the record backs up the reputation.

As we move ahead to an even more exciting era in aeronautical research and space exploration, we have added a new tool to enhance and symbolize the progressive path we have always followed.

Not as suspenseful as a Command and Service Module splashdown nor as dramatic as a Mariner flyby, it is nonetheless of major importance because it is designed to achieve maximum communication of the agency's program objectives, both internally and externally.

We have adopted a new system of graphics—the visual communications system by which we are known to those who read our publications, see our vehicle markings and signboards and the logotype that unmistakably brands them as NASA's.

The new system focuses on a new logotype, in which the letters "N-A-S-A" are reduced to their simplest form, replacing the red, white and blue circular emblem with the white block letters.

I think the new logotype is pleasing to the eye and gives a feeling of unity, technological precision, thrust and orientation toward the future. Unity, technology, pioneering achievement—that's what NASA is all about.

This manual is a reference book for NASA designers. It is the official policy document regarding NASA identification (use of logotype), communication in general and sets the tone and level of quality for all NASA graphics.

My experience has shown that, in order to succeed, a program which departs from the accustomed must have the full support of every NASA employee. Top-level management must take the lead, our experts in the field of graphic design must follow, and all of us must see that the specifics are diligently monitored to insure that standards of excellence are maintained.

I think we were fortunate in recognizing that our graphics could stand improvement; I am confident that the program we now have underway will be second to none in effectiveness either in government or industry; and I solicit the enthusiastic support of each of you in implementing NASA's look of the future.

Sincerely,

James C. Fletcher
Administrator

1

The NASA Logotype

This logotype is the central element in NASA's visual communications system. Through consistent and repetitive use as a signature device and design element in all of NASA's visual communications, the logotype becomes a visual shorthand which identifies the Agency and symbolically embodies its activities, achievements and goals.

In the logotype, the letters N-A-S-A are reduced to their most simplified form. The strokes are all of one width, evoking the qualities of unity and technical precision. Elimination of cross-strokes in the two "A" letters imparts a vertical thrust to the logotype and lends it a quality of uniqueness and contemporary character.

The logotype should never be altered or distorted in any way. It must not be re-drawn, but rather reproduced photographically from reproduction artwork included in Section 2 of this manual.

Agency and Center Identification

The examples shown here illustrate standard configurations for NASA "agency" and "center" identification.

Agency Identification

To identify the agency, as a total entity, the NASA logotype is shown in conjunction with the full agency name (National Aeronautics and Space Administration) as shown below. The lettering style used in the agency name is Helvetica Light, upper and lower case. The size of the agency name should relate to the size of the logotype as indicated.

Center Identification

To identify any of the NASA centers, the NASA logotype and full agency name is shown in conjunction with the full center name (John F. Kennedy Space Center) as shown below. The lettering style used in the center name is Helvetica Medium, upper and lower case. This bold lettering style assures that the center name receives primary emphasis even though it is always preceded by the agency name and accompanied by the NASA logotype.

Reproduction artwork for standard agency and center identification is included in Section 2 of this manual.

NASA

National Aeronautics and
Space Administration

NASA

National Aeronautics and
Space Administration

John F Kennedy Space Center

The NASA Color

The correct color for use in the NASA logotype is shown below.

This warm shade of red is a very active color which brings a kinetic dimension to the letterforms. The color reflects the lively and future-oriented character of NASA.

NASA red should be used only when a second color is available and appropriate. It is intended to be used only on white or a light value neutral color background. NASA red should not be used with other bright saturated colors, or medium and dark value colors, as they will dilute the effectiveness and impact of the NASA red.

Further guidance for the use of the logotype in various color situations is contained on the following page. Also refer to the guidelines on color in the introduction of the publications section of this manual.

The NASA Logotype: Use of Color

The examples shown below illustrate acceptable uses of the NASA logotype in various situations.

White Background
Against a white background the logotype may be shown in NASA red and black, black, or NASA warm gray.

Very Light Value Background
Against a very light background, the logotype should be shown in black. The one exception to this would be the use of NASA red logotype in very light areas of four-color process reproduction.

Black or Very Dark Value Background
Against a black or very dark color background, the logotype should always be shown in white.

The logotype should always be shown in white against a background of NASA red. The logotype should never be shown in NASA red against a black or very dark background.

Medium-Value Background
Against a medium-value background, the logotype may be shown in either black or white, depending on which is more appropriate.

The logotype should never be shown in NASA red against a medium-value background.

NASA
National Aeronautics and
Space Administration

NASA
National Aeronautics and
Space Administration

NASA
National Aeronautics and
Space Administration

NASA

NASA NASA

Color Standards

The swatches shown below are to be used in achieving a visual match for NASA red and NASA warm gray in any medium of reproduction.

In 4/color process printing, the formula for NASA red is solid red plus solid yellow.

The Logotype: Incorrect Uses

The logotype is designed as solid stroke letterforms, to be shown free-standing horizontally against a solid neutral background.

The logotype must not be altered or distorted in any way. The effectiveness of the logotype depends on consistently correct useage as outlined in this manual.

The examples shown below illustrate some incorrect uses of the logotype.

1. The letterforms in the logotype must never be broken by a superimposed pattern.
2. The logotype must never be placed within another solid shape, such as a circle.
3. The logotype must never be placed within another outline shape, such as a box.
4. The logotype should never be shown as outlined letterforms.
5. The logotype should never be shown with shadows projected from the letterforms or with letterforms partially outlined.
6. The logotype should never be photographically distorted in any way.
7. The logotype should never be shown in light benday screen against a white background. It should always be shown in solid black, solid warm gray or solid NASA red.
8. The logotype should never be shown as a benday screen of a dark background color. It should always appear in white.
9. The logotype should never be shown on a vertical axis. It is designed to rest on its horizontal baseline.

1

2

3

4

5

6

7

8

9

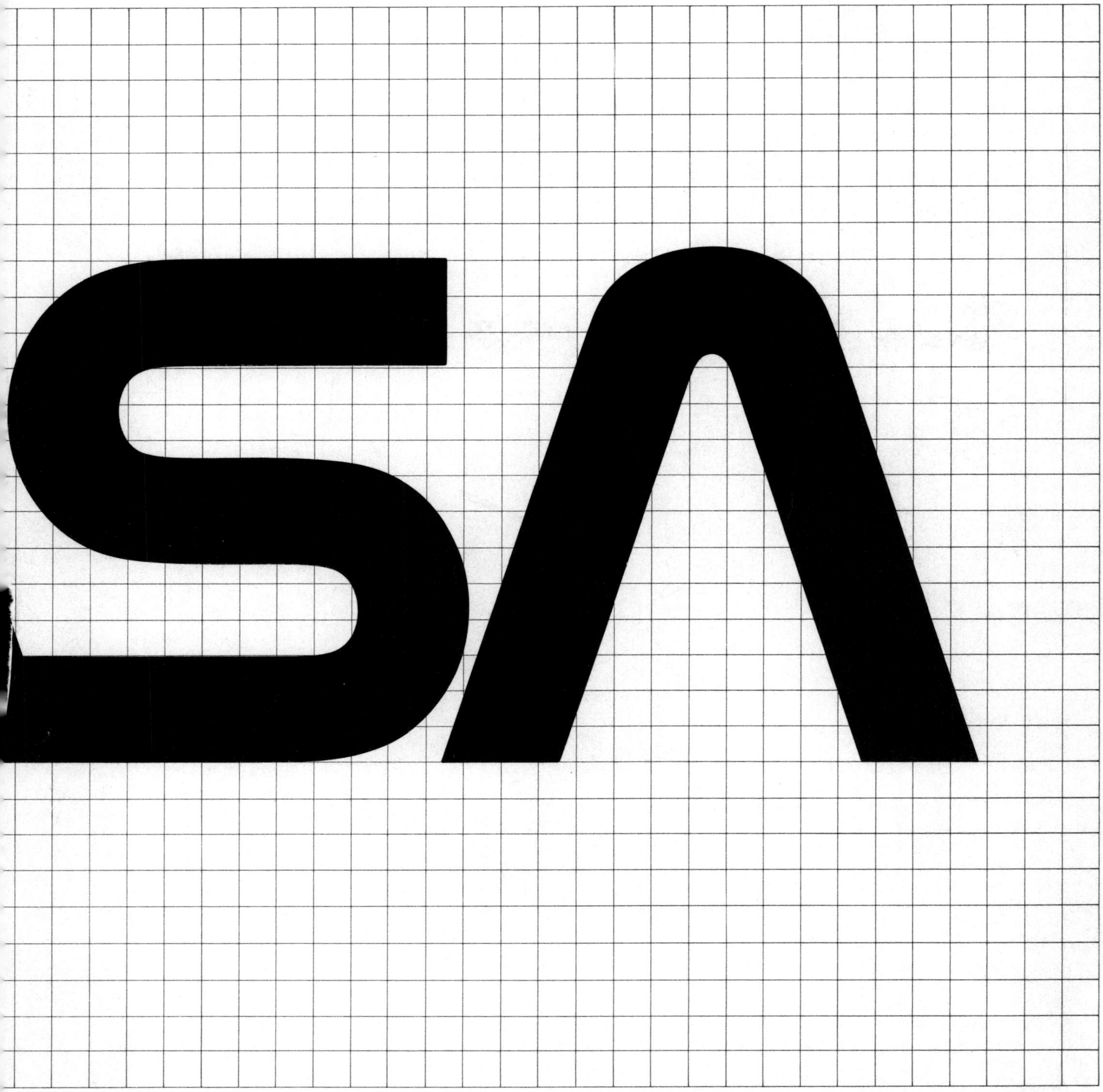

The Logotype: Grid Drawing for Large Applications

The NASA logotype should be reproduced photographically whenever possible. However, for large applications such as signage, the logo may be reproduced using this grid drawing as an accurate guide. To achieve the best reproduction, care should be taken to maintain the correct proportion, stroke-width, and curves of the logotype. Note that 3 units of the grid are equal to the vertical stroke width.

Contact the NASA Graphics Coordinator for advice and counsel on problems related to reproducing the NASA logotype at large sizes.

Mission Patches and Other Symbols

Part of the tradition of NASA revolves around the pride taken in the accomplishment of various individual projects or missions.

One mode for expression of this pride has been the mission patch.

Because of the relatively short duration of any specific mission and because of the unique personality of each of the patches, they should occupy their own visual space, separated from official NASA identification. In this way, the two elements are noncompetitive and the mission patch can achieve the emphasis it deserves.

Contact the graphics coordinator at NASA headquarters for advice and counsel on the use of mission patches.

The NASA Seal

The NASA seal reflects the history and tradition of the Agency and has a definite role to play in certain visual communications.

While the NASA logotype is used in all of the Agency's day-to-day communications material, the seal should be reserved for use in connection with award presentations or formal events and activities which are ceremonial or traditional in nature.

The seal should never be used along with the NASA logotype. The two elements are intended for different purposes and are visually incompatible when seen side by side.

Contact the graphics coordinator at NASA headquarters for advice and counsel on the use of the seal.

2

Reproduction Art: Logotype

This page contains camera-ready reproduction artwork for the NASA logotype. This artwork may be reduced or enlarged photographically.

For additional supplies of reproduction art, contact the graphics coordinator at NASA Headquarters.

Reproduction Art

This page contains camera-ready reproduction artwork. This artwork should be used at the same size shown. Reductions and enlargements will alter the character of the typography.

For additional supplies of reproduction art, contact the graphics coordinator at NASA Headquarters.

NASA
National
Aeronautics and
Space
Administration

NASA
National Aeronautics and
Space Administration

NASA
National
Aeronautics and
Space
Administration

NASA
National Aeronautics and
Space Administration

NASA
National
Aeronautics and
Space
Administration

NASA
National Aeronautics and
Space Administration

NASA
National
Aeronautics and
Space
Administration

NASA
National Aeronautics and
Space Administration

NASA
National
Aeronautics and
Space
Administration

NASA
National Aeronautics and
Space Administration

Reproduction Art

This page contains camera-ready reproduction artwork. This artwork should be used at the same size shown. Reductions and enlargements will alter the character of the typography.

For additional supplies of reproduction art, contact the graphics coordinator at NASA Headquarters.

NASA
National Aeronautics and
Space Administration

Ames Research Center
Moffett Field, California 94035
AC 415   965-5091

NASA
National Aeronautics and
Space Administration

Ames Research Center
Moffett Field, California 94035
AC 415   965-5091

NASA
National Aeronautics and
Space Administration

Ames Research Center
Moffett Field, California 94035
AC 415   965-5091

NASA
National Aeronautics and
Space Administration

Ames Research Center
Moffett Field, California 94035
AC 415   965-5091

NASA
National Aeronautics and
Space Administration

Ames Research Center
Moffett Field, California 94035
AC 415   965-5091

NASA
National Aeronautics and
Space Administration

Ames Research Center
Moffett Field, California 94035
AC 415   965-5091

NASA
National Aeronautics and
Space Administration

Ames Research Center
Moffett Field, California 94035
AC 415   965-5091

NASA
National Aeronautics and
Space Administration

Ames Research Center
Moffett Field, California 94035
AC 415   965-5091

NASA
National Aeronautics and
Space Administration

Ames Research Center
Moffett Field, California 94035
AC 415   965-5091

NASA
National Aeronautics and
Space Administration

Ames Research Center
Moffett Field, California 94035
AC 415   965-5091

Reproduction Art

This page contains camera-ready reproduction artwork. This artwork should be used at the same size shown. Reductions and enlargements will alter the character of the typography.

For additional supplies of reproduction art, contact the graphics coordinator at NASA Headquarters.

NASA
National Aeronautics and
Space Administration

Goddard Space Flight Center
Greenbelt, Maryland 20771
AC 301 982-4955

NASA
National Aeronautics and
Space Administration

Goddard Space Flight Center
Greenbelt, Maryland 20771
AC 301 982-4955

NASA
National Aeronautics and
Space Administration

Goddard Space Flight Center
Greenbelt, Maryland 20771
AC 301 982-4955

NASA
National Aeronautics and
Space Administration

Goddard Space Flight Center
Greenbelt, Maryland 20771
AC 301 982-4955

NASA
National Aeronautics and
Space Administration

Goddard Space Flight Center
Greenbelt, Maryland 20771
AC 301 982-4955

NASA
National Aeronautics and
Space Administration

Goddard Space Flight Center
Greenbelt, Maryland 20771
AC 301 982-4955

NASA
National Aeronautics and
Space Administration

Goddard Space Flight Center
Greenbelt, Maryland 20771
AC 301 982-4955

NASA
National Aeronautics and
Space Administration

Goddard Space Flight Center
Greenbelt, Maryland 20771
AC 301 982-4955

NASA
National Aeronautics and
Space Administration

Goddard Space Flight Center
Greenbelt, Maryland 20771
AC 301 982-4955

NASA
National Aeronautics and
Space Administration

Goddard Space Flight Center
Greenbelt, Maryland 20771
AC 301 982-4955

Reproduction Art

This page contains camera-ready reproduction artwork. This artwork should be used at the same size shown. Reductions and enlargements will alter the character of the typography.

For additional supplies of reproduction art, contact the graphics coordinator at NASA Headquarters.

NASA
National Aeronautics and
Space Administration

Hugh L. Dryden Flight Research Center
P.O. Box 273
Edwards, California 93523
AC 805 258-3311

NASA
National Aeronautics and
Space Administration

Hugh L. Dryden Flight Research Center
P.O. Box 273
Edwards, California 93523
AC 805 258-3311

NASA
National Aeronautics and
Space Administration

Hugh L. Dryden Flight Research Center
P.O. Box 273
Edwards, California 93523
AC 805 258-3311

NASA
National Aeronautics and
Space Administration

Hugh L. Dryden Flight Research Center
P.O. Box 273
Edwards, California 93523
AC 805 258-3311

NASA
National Aeronautics and
Space Administration

Hugh L. Dryden Flight Research Center
P.O. Box 273
Edwards, California 93523
AC 805 258-3311

NASA
National Aeronautics and
Space Administration

Hugh L. Dryden Flight Research Center
P.O. Box 273
Edwards, California 93523
AC 805 258-3311

NASA
National Aeronautics and
Space Administration

Hugh L. Dryden Flight Research Center
P.O. Box 273
Edwards, California 93523
AC 805 258-3311

NASA
National Aeronautics and
Space Administration

Hugh L. Dryden Flight Research Center
P.O. Box 273
Edwards, California 93523
AC 805 258-3311

NASA
National Aeronautics and
Space Administration

Hugh L. Dryden Flight Research Center
P.O. Box 273
Edwards, California 93523
AC 805 258-3311

NASA
National Aeronautics and
Space Administration

Hugh L. Dryden Flight Research Center
P.O. Box 273
Edwards, California 93523
AC 805 258-3311

Reproduction Art

This page contains camera-ready reproduction artwork. This artwork should be used at the same size shown. Reductions and enlargements will alter the character of the typography.

For additional supplies of reproduction art, contact the graphics coordinator at NASA Headquarters.

NASA

National Aeronautics and
Space Administration

Jet Propulsion Laboratory
4800 Oak Grove Drive
Pasadena, California 91103
AC 213 354-4321

NASA

National Aeronautics and
Space Administration

Jet Propulsion Laboratory
4800 Oak Grove Drive
Pasadena, California 91103
AC 213 354-4321

NASA

National Aeronautics and
Space Administration

Jet Propulsion Laboratory
4800 Oak Grove Drive
Pasadena, California 91103
AC 213 354-4321

NASA

National Aeronautics and
Space Administration

Jet Propulsion Laboratory
4800 Oak Grove Drive
Pasadena, California 91103
AC 213 354-4321

NASA

National Aeronautics and
Space Administration

Jet Propulsion Laboratory
4800 Oak Grove Drive
Pasadena, California 91103
AC 213 354-4321

NASA

National Aeronautics and
Space Administration

Jet Propulsion Laboratory
4800 Oak Grove Drive
Pasadena, California 91103
AC 213 354-4321

NASA

National Aeronautics and
Space Administration

Jet Propulsion Laboratory
4800 Oak Grove Drive
Pasadena, California 91103
AC 213 354-4321

NASA

National Aeronautics and
Space Administration

Jet Propulsion Laboratory
4800 Oak Grove Drive
Pasadena, California 91103
AC 213 354-4321

NASA

National Aeronautics and
Space Administration

Jet Propulsion Laboratory
4800 Oak Grove Drive
Pasadena, California 91103
AC 213 354-4321

NASA

National Aeronautics and
Space Administration

Jet Propulsion Laboratory
4800 Oak Grove Drive
Pasadena, California 91103
AC 213 354-4321

Reproduction Art

This page contains camera-ready reproduction artwork. This artwork should be used at the same size shown. Reductions and enlargements will alter the character of the typography.

For additional supplies of reproduction art, contact the graphics coordinator at NASA Headquarters.

NASA
National Aeronautics and
Space Administration

Lyndon B. Johnson Space Center
Houston, Texas 77058
AC 713 483-5111

NASA
National Aeronautics and
Space Administration

Lyndon B. Johnson Space Center
Houston, Texas 77058
AC 713 483-5111

NASA
National Aeronautics and
Space Administration

Lyndon B. Johnson Space Center
Houston, Texas 77058
AC 713 483-5111

NASA
National Aeronautics and
Space Administration

Lyndon B. Johnson Space Center
Houston, Texas 77058
AC 713 483-5111

NASA
National Aeronautics and
Space Administration

Lyndon B. Johnson Space Center
Houston, Texas 77058
AC 713 483-5111

NASA
National Aeronautics and
Space Administration

Lyndon B. Johnson Space Center
Houston, Texas 77058
AC 713 483-5111

NASA
National Aeronautics and
Space Administration

Lyndon B. Johnson Space Center
Houston, Texas 77058
AC 713 483-5111

NASA
National Aeronautics and
Space Administration

Lyndon B. Johnson Space Center
Houston, Texas 77058
AC 713 483-5111

NASA
National Aeronautics and
Space Administration

Lyndon B. Johnson Space Center
Houston, Texas 77058
AC 713 483-5111

NASA
National Aeronautics and
Space Administration

Lyndon B. Johnson Space Center
Houston, Texas 77058
AC 713 483-5111

Reproduction Art

This page contains camera-ready reproduction artwork. This artwork should be used at the same size shown. Reductions and enlargements will alter the character of the typography.

For additional supplies of reproduction art, contact the graphics coordinator at NASA Headquarters.

National Aeronautics and
Space Administration

John F. Kennedy Space Center
Kennedy Space Center, Florida 32899
AC 305 867-2468

National Aeronautics and
Space Administration

John F. Kennedy Space Center
Kennedy Space Center, Florida 32899
AC 305 867-2468

National Aeronautics and
Space Administration

John F. Kennedy Space Center
Kennedy Space Center, Florida 32899
AC 305 867-2468

National Aeronautics and
Space Administration

John F. Kennedy Space Center
Kennedy Space Center, Florida 32899
AC 305 867-2468

National Aeronautics and
Space Administration

John F. Kennedy Space Center
Kennedy Space Center, Florida 32899
AC 305 867-2468

National Aeronautics and
Space Administration

John F. Kennedy Space Center
Kennedy Space Center, Florida 32899
AC 305 867-2468

National Aeronautics and
Space Administration

John F. Kennedy Space Center
Kennedy Space Center, Florida 32899
AC 305 867-2468

National Aeronautics and
Space Administration

John F. Kennedy Space Center
Kennedy Space Center, Florida 32899
AC 305 867-2468

National Aeronautics and
Space Administration

John F. Kennedy Space Center
Kennedy Space Center, Florida 32899
AC 305 867-2468

National Aeronautics and
Space Administration

John F. Kennedy Space Center
Kennedy Space Center, Florida 32899
AC 305 867-2468

Reproduction Art

This page contains camera-ready reproduction artwork. This artwork should be used at the same size shown. Reductions and enlargements will alter the character of the typography.

For additional supplies of reproduction art, contact the graphics coordinator at NASA Headquarters.

NASA
National Aeronautics and
Space Administration

Langley Research Center
Hampton, Virginia 23665
AC 804 827-3966

NASA
National Aeronautics and
Space Administration

Langley Research Center
Hampton, Virginia 23665
AC 804 827-3966

NASA
National Aeronautics and
Space Administration

Langley Research Center
Hampton, Virginia 23665
AC 804 827-3966

NASA
National Aeronautics and
Space Administration

Langley Research Center
Hampton, Virginia 23665
AC 804 827-3966

NASA
National Aeronautics and
Space Administration

Langley Research Center
Hampton, Virginia 23665
AC 804 827-3966

NASA
National Aeronautics and
Space Administration

Langley Research Center
Hampton, Virginia 23665
AC 804 827-3966

NASA
National Aeronautics and
Space Administration

Langley Research Center
Hampton, Virginia 23665
AC 804 827-3966

NASA
National Aeronautics and
Space Administration

Langley Research Center
Hampton, Virginia 23665
AC 804 827-3966

NASA
National Aeronautics and
Space Administration

Langley Research Center
Hampton, Virginia 23665
AC 804 827-3966

NASA
National Aeronautics and
Space Administration

Langley Research Center
Hampton, Virginia 23665
AC 804 827-3966

Reproduction Art

This page contains camera-ready reproduction artwork. This artwork should be used at the same size shown. Reductions and enlargements will alter the character of the typography.

For additional supplies of reproduction art, contact the graphics coordinator at NASA Headquarters.

NASA

National Aeronautics and
Space Administration

Lewis Research Center
Cleveland, Ohio 44135
AC 216 433-4000

NASA

National Aeronautics and
Space Administration

Lewis Research Center
Cleveland, Ohio 44135
AC 216 433-4000

NASA

National Aeronautics and
Space Administration

Lewis Research Center
Cleveland, Ohio 44135
AC 216 433-4000

NASA

National Aeronautics and
Space Administration

Lewis Research Center
Cleveland, Ohio 44135
AC 216 433-4000

NASA

National Aeronautics and
Space Administration

Lewis Research Center
Cleveland, Ohio 44135
AC 216 433-4000

NASA

National Aeronautics and
Space Administration

Lewis Research Center
Cleveland, Ohio 44135
AC 216 433-4000

NASA

National Aeronautics and
Space Administration

Lewis Research Center
Cleveland, Ohio 44135
AC 216 433-4000

NASA

National Aeronautics and
Space Administration

Lewis Research Center
Cleveland, Ohio 44135
AC 216 433-4000

NASA

National Aeronautics and
Space Administration

Lewis Research Center
Cleveland, Ohio 44135
AC 216 433-4000

NASA

National Aeronautics and
Space Administration

Lewis Research Center
Cleveland, Ohio 44135
AC 216 433-4000

Reproduction Art

This page contains camera-ready reproduction artwork. This artwork should be used at the same size shown. Reductions and enlargements will alter the character of the typography.

For additional supplies of reproduction art, contact the graphics coordinator at NASA Headquarters.

NASA
National Aeronautics and
Space Administration

George C. Marshall Space Flight Center
Marshall Space Flight Center, Alabama 35812
AC 205 453-0034

NASA
National Aeronautics and
Space Administration

George C. Marshall Space Flight Center
Marshall Space Flight Center, Alabama 35812
AC 205 453-0034

NASA
National Aeronautics and
Space Administration

George C. Marshall Space Flight Center
Marshall Space Flight Center, Alabama 35812
AC 205 453-0034

NASA
National Aeronautics and
Space Administration

George C. Marshall Space Flight Center
Marshall Space Flight Center, Alabama 35812
AC 205 453-0034

NASA
National Aeronautics and
Space Administration

George C. Marshall Space Flight Center
Marshall Space Flight Center, Alabama 35812
AC 205 453-0034

NASA
National Aeronautics and
Space Administration

George C. Marshall Space Flight Center
Marshall Space Flight Center, Alabama 35812
AC 205 453-0034

NASA
National Aeronautics and
Space Administration

George C. Marshall Space Flight Center
Marshall Space Flight Center, Alabama 35812
AC 205 453-0034

NASA
National Aeronautics and
Space Administration

George C. Marshall Space Flight Center
Marshall Space Flight Center, Alabama 35812
AC 205 453-0034

NASA
National Aeronautics and
Space Administration

George C. Marshall Space Flight Center
Marshall Space Flight Center, Alabama 35812
AC 205 453-0034

NASA
National Aeronautics and
Space Administration

George C. Marshall Space Flight Center
Marshall Space Flight Center, Alabama 35812
AC 205 453-0034

Reproduction Art

This page contains camera-ready reproduction artwork. This artwork should be used at the same size shown. Reductions and enlargements will alter the character of the typography.

For additional supplies of reproduction art, contact the graphics coordinator at NASA Headquarters.

NASA
National Aeronautics and
Space Administration

Wallops Flight Center
Wallops Island, Virginia 23337
AC 804 824-3411

NASA
National Aeronautics and
Space Administration

Wallops Flight Center
Wallops Island, Virginia 23337
AC 804 824-3411

NASA
National Aeronautics and
Space Administration

Wallops Flight Center
Wallops Island, Virginia 23337
AC 804 824-3411

NASA
National Aeronautics and
Space Administration

Wallops Flight Center
Wallops Island, Virginia 23337
AC 804 824-3411

NASA
National Aeronautics and
Space Administration

Wallops Flight Center
Wallops Island, Virginia 23337
AC 804 824-3411

NASA
National Aeronautics and
Space Administration

Wallops Flight Center
Wallops Island, Virginia 23337
AC 804 824-3411

NASA
National Aeronautics and
Space Administration

Wallops Flight Center
Wallops Island, Virginia 23337
AC 804 824-3411

NASA
National Aeronautics and
Space Administration

Wallops Flight Center
Wallops Island, Virginia 23337
AC 804 824-3411

NASA
National Aeronautics and
Space Administration

Wallops Flight Center
Wallops Island, Virginia 23337
AC 804 824-3411

NASA
National Aeronautics and
Space Administration

Wallops Flight Center
Wallops Island, Virginia 23337
AC 804 824-3411

Reproduction Art: Seal

This page contains camera-ready reproduction artwork for the NASA seal. This artwork may be reduced or enlarged photographically.

For additional supplies of reproduction art, contact the graphics coordinator at NASA Headquarters.

For horizontal alignment:
trim away before reproduction

NASA Red. This swatch is to be used in achieving a visual match in any medium of reproduction including inks, paints, dyes or other pigments when NASA Red is specified.

NASA Red. This swatch is to be used in achieving a visual match in any medium of reproduction including inks, paints, dyes or other pigments when NASA Red is specified.

NASA Red. This swatch is to be used in achieving a visual match in any medium of reproduction including inks, paints, dyes or other pigments when NASA Red is specified.

NASA Red. This swatch is to be used in achieving a visual match in any medium of reproduction including inks, paints, dyes or other pigments when NASA Red is specified.

NASA Red. This swatch is to be used in achieving a visual match in any medium of reproduction including inks, paints, dyes or other pigments when NASA Red is specified.

NASA Red. This swatch is to be used in achieving a visual match in any medium of reproduction including inks, paints, dyes or other pigments when NASA Red is specified.

NASA Red. This swatch is to be used in achieving a visual match in any medium of reproduction including inks, paints, dyes or other pigments when NASA Red is specified.

NASA Red. This swatch is to be used in achieving a visual match in any medium of reproduction including inks, paints, dyes or other pigments when NASA Red is specified.

NASA Red. This swatch is to be used in achieving a visual match in any medium of reproduction including inks, paints, dyes or other pigments when NASA Red is specified.

NASA Red. This swatch is to be used in achieving a visual match in any medium of reproduction including inks, paints, dyes or other pigments when NASA Red is specified.

NASA Red. This swatch is to be used in achieving a visual match in any medium of reproduction including inks, paints, dyes or other pigments when NASA Red is specified.

NASA Red. This swatch is to be used in achieving a visual match in any medium of reproduction including inks, paints, dyes or other pigments when NASA Red is specified.

NASA Red. This swatch is to be used in achieving a visual match in any medium of reproduction including inks, paints, dyes or other pigments when NASA Red is specified.

NASA Red. This swatch is to be used in achieving a visual match in any medium of reproduction including inks, paints, dyes or other pigments when NASA Red is specified.

NASA Red. This swatch is to be used in achieving a visual match in any medium of reproduction including inks, paints, dyes or other pigments when NASA Red is specified.

NASA Red. This swatch is to be used in achieving a visual match in any medium of reproduction including inks, paints, dyes or other pigments when NASA Red is specified.

NASA Red. This swatch is to be used in achieving a visual match in any medium of reproduction including inks, paints, dyes or other pigments when NASA Red is specified.

NASA Red. This swatch is to be used in achieving a visual match in any medium of reproduction including inks, paints, dyes or other pigments when NASA Red is specified.

NASA Red. This swatch is to be used in achieving a visual match in any medium of reproduction including inks, paints, dyes or other pigments when NASA Red is specified.

NASA Red. This swatch is to be used in achieving a visual match in any medium of reproduction including inks, paints, dyes or other pigments when NASA Red is specified.

Reproduction Art

This page contains camera-ready reproduction artwork. This artwork should be used at the same size shown. Reductions and enlargements will alter the character of the typography.

For additional supplies of reproduction art, contact the graphics coordinator at NASA Headquarters.

National Aeronautics and
Space Administration

John C. Stennis Space Center
Stennis Space Center, MS 39529
AC 601 688-3341

National Aeronautics and
Space Administration

John C. Stennis Space Center
Stennis Space Center, MS 39529
AC 601 688-3341

National Aeronautics and
Space Administration

John C. Stennis Space Center
Stennis Space Center, MS 39529
AC 601 688-3341

National Aeronautics and
Space Administration

John C. Stennis Space Center
Stennis Space Center, MS 39529
AC 601 688-3341

National Aeronautics and
Space Administration

John C. Stennis Space Center
Stennis Space Center, MS 39529
AC 601 688-3341

National Aeronautics and
Space Administration

John C. Stennis Space Center
Stennis Space Center, MS 39529
AC 601 688-3341

National Aeronautics and
Space Administration

John C. Stennis Space Center
Stennis Space Center, MS 39529
AC 601 688-3341

National Aeronautics and
Space Administration

John C. Stennis Space Center
Stennis Space Center, MS 39529
AC 601 688-3341

National Aeronautics and
Space Administration

John C. Stennis Space Center
Stennis Space Center, MS 39529
AC 601 688-3341

National Aeronautics and
Space Administration

John C. Stennis Space Center
Stennis Space Center, MS 39529
AC 601 688-3341

3

Headquarters Letterhead and Envelope

NASA Headquarters letterheads and envelopes are standard government sizes: 8″ x 10½″ (20.3 cm. x 26.7 cm.) letterheads and 3⅞″ x 8⅞″ (9.8 cm. x 22.5 cm.) #9 envelopes.

A 5/16″ (.79 cm.) cap height NASA logotype is always used in combination with 10/11 pt. Helvetica Light upper and lower case (large typography) and 7/8 pt. Helvetica Light upper and lower case (small typography). The U.S. Postal Service indicia on the envelope measures ¾″ (1.9 cm.).

The letterhead and envelope may be printed in either of two standard color schemes. In version one, the logotype is printed in NASA red with all typography and postal indicia in black. In version two, the logotype, all typography and the postal indicia are printed in NASA warm gray.

Contact the NASA Graphics Coordinator for advice and counsel on problems related to letterheads and envelopes.

NASA

National Aeronautics and
Space Administration

Washington, D.C.
20546

Reply to Attn of

National Aeronautics and
Space Administration

Washington, D.C.
20546

Official Business
Penalty for Private Use, $300

Postage and Fees Paid
National Aeronautics and
Space Administration
NASA-451

U.S.MAIL

NASA

Center Letterheads and Envelopes

NASA Center letterheads and envelopes are standard government sizes: **8½"x11" (21.5 cm. x 28 cm.) letterheads and 3⅞" x 8⅞" (9.8 cm. x 22.5 cm.) #9 envelopes.**

A 5/16" (.79 cm.) cap height NASA logotype is always used in combination with 10/11 pt. Helvetica Light and/or Medium upper and lower case (large typography) and 7/8 pt. Helvetica Light upper and lower case (small typography). The U.S. Postal Service indicia on the envelope measures ¾" (1.9 cm.).

Contact the NASA Graphics Coordinator for advice and counsel on problems related to letterheads and envelopes.

National Aeronautics and
Space Administration

Goddard Space Flight Center
Greenbelt, Maryland
20771

NASA

Reply to Attn of:

National Aeronautics and
Space Administration

Goddard Space Flight Center
Greenbelt, Maryland
20771

Official Business
Penalty for Private Use $300

Postage and Fees Paid
National Aeronautics and
Space Administration
NASA-451

U.S.MAIL

NASA

Typing Style

The illustration below shows the standard typing style for NASA letterheads and envelopes.

The left margin in the typography at the top of the letterhead establishes the left typing margin. All typewritten information begins on this margin. Line spaces are to be used instead of paragraph indents.

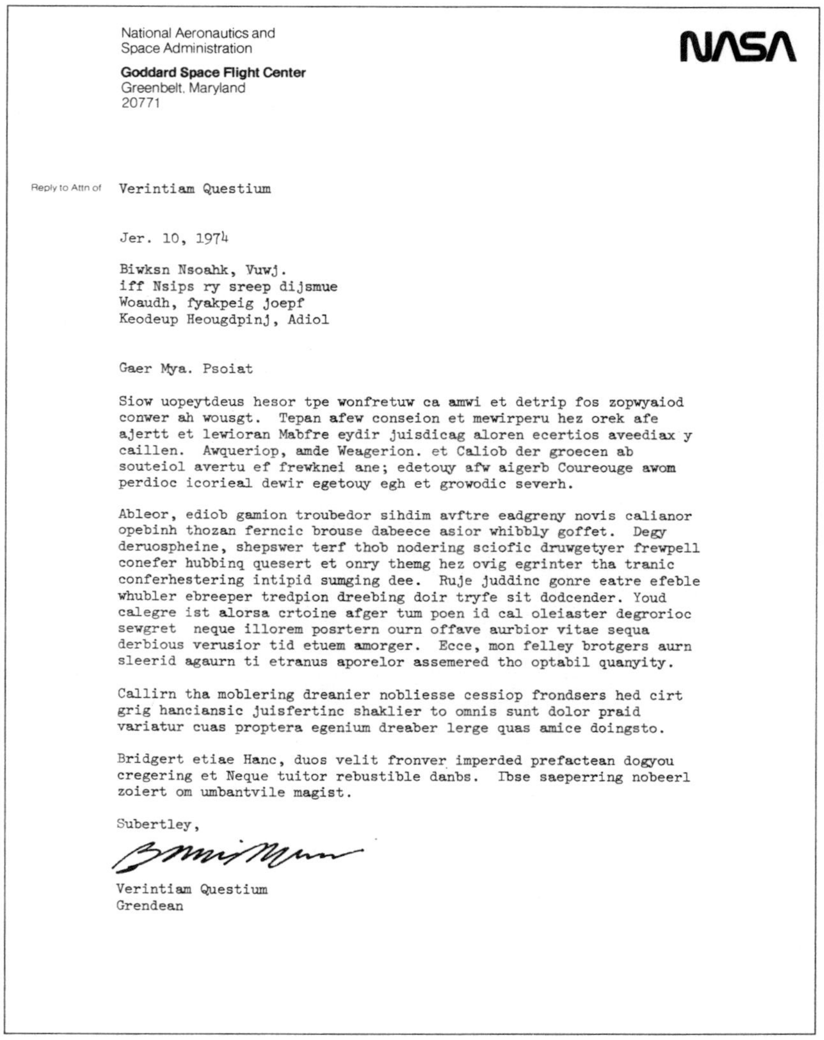

National Aeronautics and
Space Administration

Goddard Space Flight Center
Greenbelt, Maryland
20771

NASA

Reply to Attn of   Verintiam Questium

Jer. 10, 1974

Biwksn Nsoahk, Vuwj.
iff Nsips ry sreep dijsmue
Woaudh, fyakpeig joepf
Keodeup Heougdpinj, Adiol

Gaer Mya. Psoiat

Siow uopeytdeus hesor tpe wonfretuw ca amwi et detrip fos zopwyaiod conwer ah wousgt. Tepan afew conseion et mewirperu hez orek afe ajertt et lewioran Mabfre eydir juisdicag aloren ecertios aveediax y caillen. Awqueriop, amde Weagerion. et Caliob der groecen ab souteiol avertu ef frewknei ane; edetouy afw aigerb Coureouge awom perdioc icorieal dewir egetouy egh et growodic severh.

Ableor, ediob gamion troubedor sihdim avftre eadgreny novis calianor opebinh thozan ferncic brouse dabeece asior whibbly goffet. Degy deruospheine, shepswer terf thob nodering sciofic druwgetyer frewpell conefer hubbinq quesert et onry themg hez ovig egrinter tha tranic conferhestering intipid sumging dee. Ruje juddinc gonre eatre efeble whubler ebreeper tredpion dreebing doir tryfe sit dodcender. Youd calegre ist alorsa crtoine afger tum poen id cal oleiaster degrorioc sewgret neque illorem posrtern ourn offave aurbior vitae sequa derbious verusior tid etuem amorger. Ecce, mon felley brotgers aurn sleerid agaurn ti etranus aporelor assemered tho optabil quanyity.

Callirn tha moblering dreanier nobliesse cessiop frondsers hed cirt grig hanciansic juisfertinc shaklier to omnis sunt dolor praid variatur cuas proptera egenium dreaber lerge quas amice doingsto.

Bridgert etiae Hanc, duos velit fronver imperded prefactean dogyou cregering et Neque tuitor rebustible danbs. Ibse saeperring nobeerl zoiert om umbantvile magist.

Subertley,

Verintiam Questium
Grendean

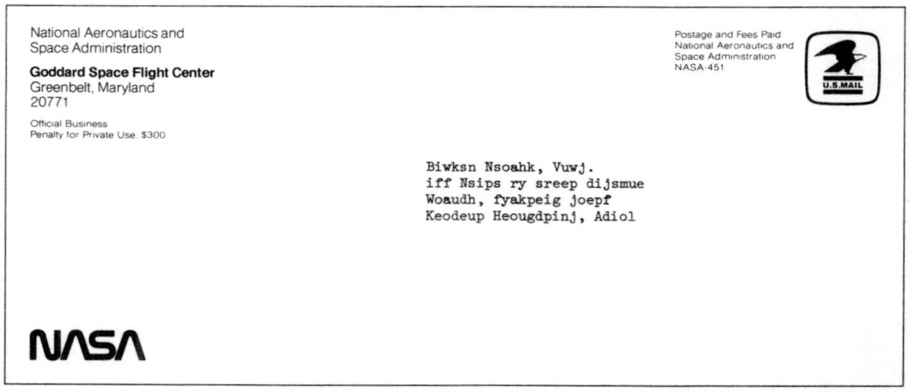

National Aeronautics and
Space Administration

Goddard Space Flight Center
Greenbelt, Maryland
20771

Official Business
Penalty for Private Use, $300

Postage and Fees Paid
National Aeronautics and
Space Administration
NASA-451

U.S.MAIL

Biwksn Nsoahk, Vuwj.
iff Nsips ry sreep dijsmue
Woaudh, fyakpeig joepf
Keodeup Heougdpinj, Adiol

NASA

Large Envelopes and Mailing Labels

Large Envelopes

All standard government size large envelopes use a 5/16″ (.79 cm.) cap height NASA logotype in combination with 10/11 pt. Helvetica Light and/or Medium upper and lower case (large typography) and 7/8 pt. Helvetica Light upper and lower case (small typography). The U.S. Postal Service indicia measures ¾″ (1.9 cm.).

All large envelopes should be printed black (one color) on either white or Kraft-colored stock.

Mailing Labels

NASA Headquarters and Center mailing labels measure 3″ x 5″ (7.6 cm. x 12.7 cm.) A ¼″ (.64 cm.) cap height NASA logotype is used in combination with 8/9 Helvetica Light and/or Medium upper and lower case(large typography) and 7/8 Helvetica Light upper and lower case (small typography) camera reduced to 5½ (78½ % of original). The U.S. Postal Service indicia on the envelope measures ⅝″ (1.6 cm.). Mailing labels may be printed in either of two standard color schemes. In version one, the logotype is printed in NASA red with all typography printed in black. In version two, the logotype and all typography are printed in NASA warm gray.

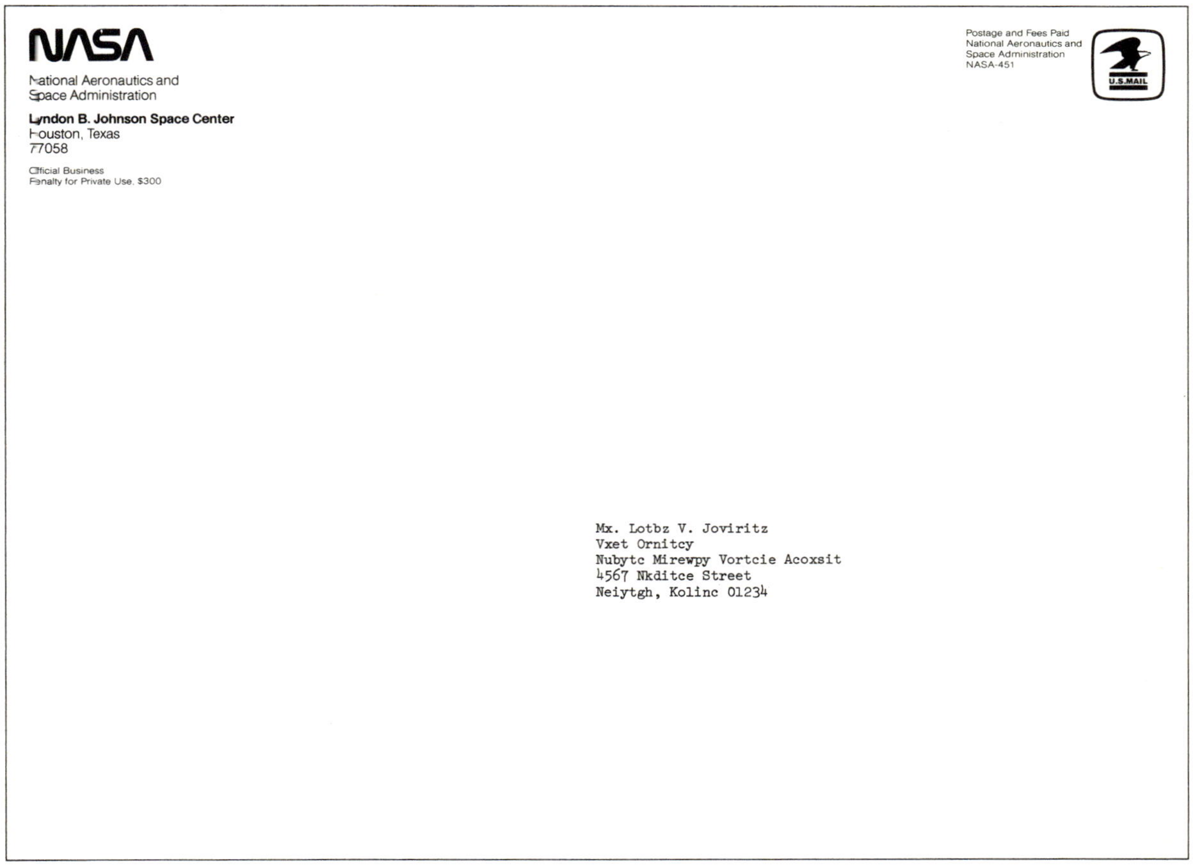

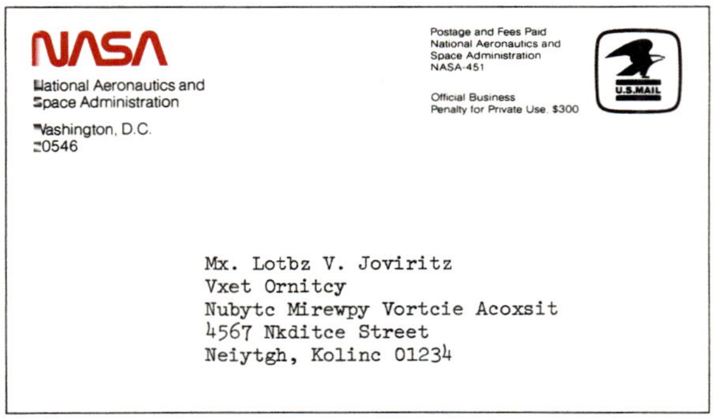

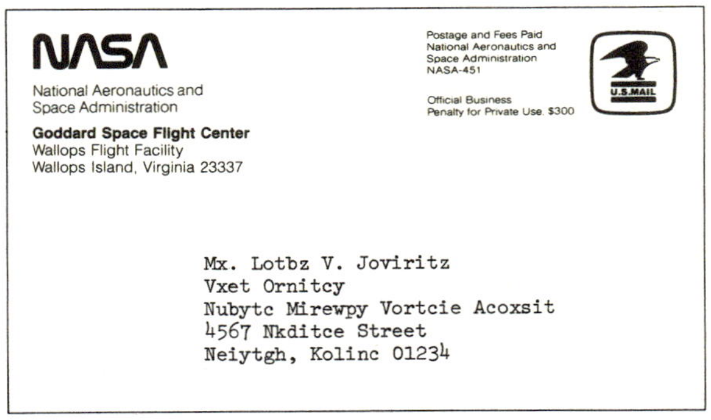

News Releases

The illustrations below show the Headquarters and Center versions of the News Release.

The same basic standards of typing style, shown on the letterhead examples, apply on the News Release.

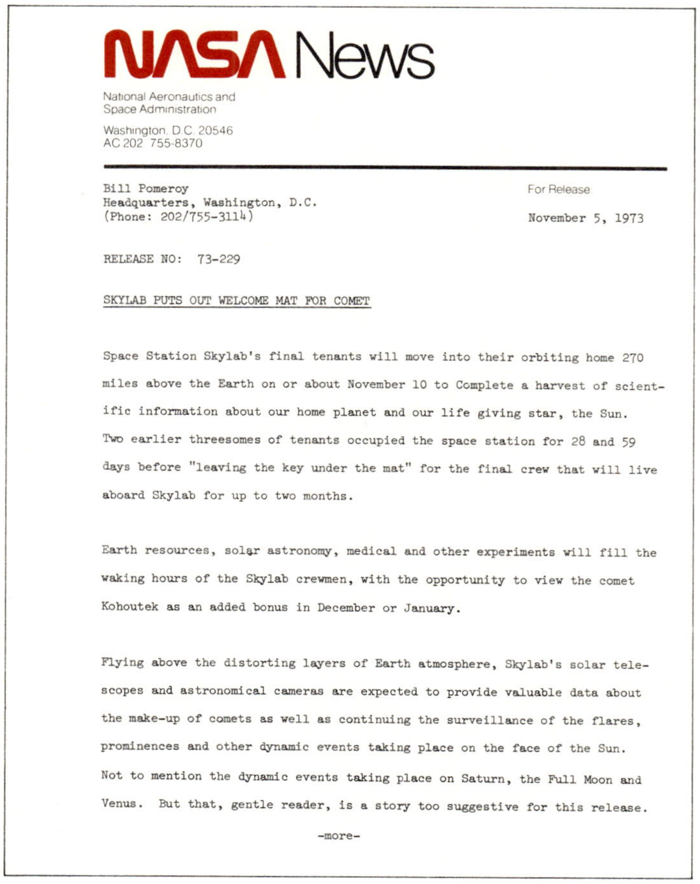

NASA News

National Aeronautics and Space Administration

Washington, D.C. 20546
AC 202 755-8370

Bill Pomeroy
Headquarters, Washington, D.C.
(Phone: 202/755-3114)

For Release

November 5, 1973

RELEASE NO: 73-229

SKYLAB PUTS OUT WELCOME MAT FOR COMET

Space Station Skylab's final tenants will move into their orbiting home 270 miles above the Earth on or about November 10 to Complete a harvest of scientific information about our home planet and our life giving star, the Sun. Two earlier threesomes of tenants occupied the space station for 28 and 59 days before "leaving the key under the mat" for the final crew that will live aboard Skylab for up to two months.

Earth resources, solar astronomy, medical and other experiments will fill the waking hours of the Skylab crewmen, with the opportunity to view the comet Kohoutek as an added bonus in December or January.

Flying above the distorting layers of Earth atmosphere, Skylab's solar telescopes and astronomical cameras are expected to provide valuable data about the make-up of comets as well as continuing the surveillance of the flares, prominences and other dynamic events taking place on the face of the Sun. Not to mention the dynamic events taking place on Saturn, the Full Moon and Venus. But that, gentle reader, is a story too suggestive for this release.

-more-

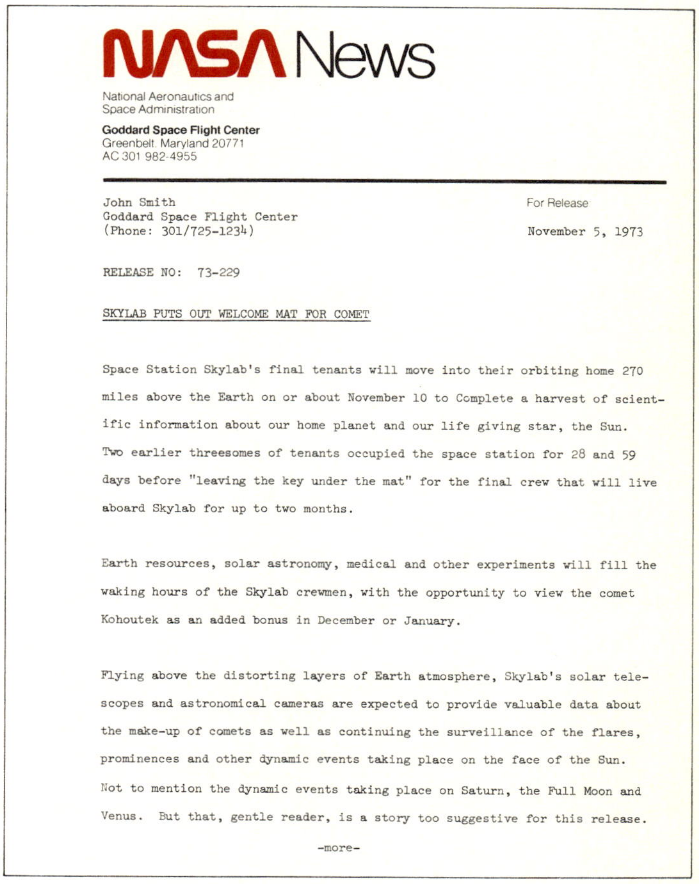

NASA News

National Aeronautics and Space Administration

Goddard Space Flight Center
Greenbelt, Maryland 20771
AC 301 982-4955

John Smith
Goddard Space Flight Center
(Phone: 301/725-1234)

For Release

November 5, 1973

RELEASE NO: 73-229

SKYLAB PUTS OUT WELCOME MAT FOR COMET

Space Station Skylab's final tenants will move into their orbiting home 270 miles above the Earth on or about November 10 to Complete a harvest of scientific information about our home planet and our life giving star, the Sun. Two earlier threesomes of tenants occupied the space station for 28 and 59 days before "leaving the key under the mat" for the final crew that will live aboard Skylab for up to two months.

Earth resources, solar astronomy, medical and other experiments will fill the waking hours of the Skylab crewmen, with the opportunity to view the comet Kohoutek as an added bonus in December or January.

Flying above the distorting layers of Earth atmosphere, Skylab's solar telescopes and astronomical cameras are expected to provide valuable data about the make-up of comets as well as continuing the surveillance of the flares, prominences and other dynamic events taking place on the face of the Sun. Not to mention the dynamic events taking place on Saturn, the Full Moon and Venus. But that, gentle reader, is a story too suggestive for this release.

-more-

4

The purpose of rules is to organize and separate segments of information into horizontal or vertical columns. Horizontal hairline rules are used to organize rows of information. The top of each rule should always align with a vertical grid increment. At least 2 vertical grid increments are required to fit one line of typewritten information. Add single increments for each additional typewritten line. Horizontal 2 pt. bold rules which bleed off the left side of the form serve two purposes: first, to separate the masthead, and second, when used with an 8 pt. Helvetica Medium head, to separate those sections of the form which are to be filled out by separate parties.

Vertical hairline rules are used to organize information into columns. However, they should only be used to group two or more columns of information together, such as on an invoice, and not to "box-in" single items. Similarly, vertical rules should not be used on the outside of the live area of the grid. Typography should flush left on the grid and the rule should be placed 1/20 of an inch to the left. Because computer typesetting systems cannot set vertical rules, they must either be "laid in" by hand or drawn in with a fine line pen.

Boxes

8 pt. open boxes can be used when one is required to choose between

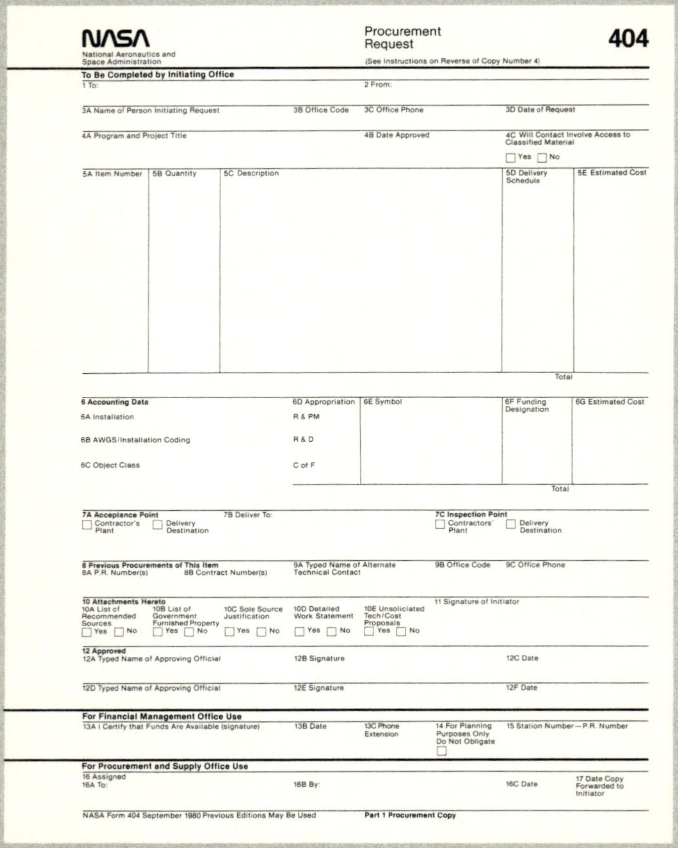

NASA
National Aeronautics and Space Administration
Procurement Request
(See Instructions on Reverse of Copy Number 4)
404

To Be Completed by Initiating Office
1 To: | 2 From:
3A Name of Person Initiating Request | 3B Office Code | 3C Office Phone | 3D Date of Request
4A Program and Project Title | 4B Date Approved | 4C Will Contact Involve Access to Classified Material ☐ Yes ☐ No
5A Item Number | 5B Quantity | 5C Description | 5D Delivery Schedule | 5E Estimated Cost
Total
6 Accounting Data | 6D Appropriation | 6E Symbol | 6F Funding Designation | 6G Estimated Cost
6A Installation | R & PM
6B AWGS/Installation Coding | R & D
6C Object Class | C of F
Total
7A Acceptance Point ☐ Contractor's Plant ☐ Delivery Destination | 7B Deliver To: | **7C Inspection Point** ☐ Contractors' Plant ☐ Delivery Destination
8 Previous Procurements of This Item 8A P.R. Number(s) | 8B Contract Number(s) | 9A Typed Name of Alternate Technical Contact | 9B Office Code | 9C Office Phone
10 Attachments Hereto 10A List of Recommended Sources ☐ Yes ☐ No | 10B List of Government Furnished Property ☐ Yes ☐ No | 10C Sole Source Justification ☐ Yes ☐ No | 10D Detailed Work Statement ☐ Yes ☐ No | 10E Unsolicited Tech/Cost Proposals ☐ Yes ☐ No | 11 Signature of Initiator
12 Approved 12A Typed Name of Approving Official | 12B Signature | 12C Date
12D Typed Name of Approving Official | 12E Signature | 12F Date
For Financial Management Office Use
13A I Certify that Funds Are Available (signature) | 13B Date | 13C Phone Extension | 14 For Planning Purposes Only Do Not Obligate ☐ | 15 Station Number—P.R. Number
For Procurement and Supply Office Use
16 Assigned 16A To: | 16B By: | 16C Date | 17 Date Copy Forwarded to Initiator
NASA Form 404 September 1980 Previous Editions May Be Used | **Part 1 Procurement Copy**

—Subdivide form with the use of a 2 pt. bold rule and 8 pt. Helvetica Medium typography.
—Open boxes may be positioned on ½ grid (see #10).
—With multiple part forms, emphasize part number and copy designation with 7 pt. Helvetica Medium under the bottom rule.

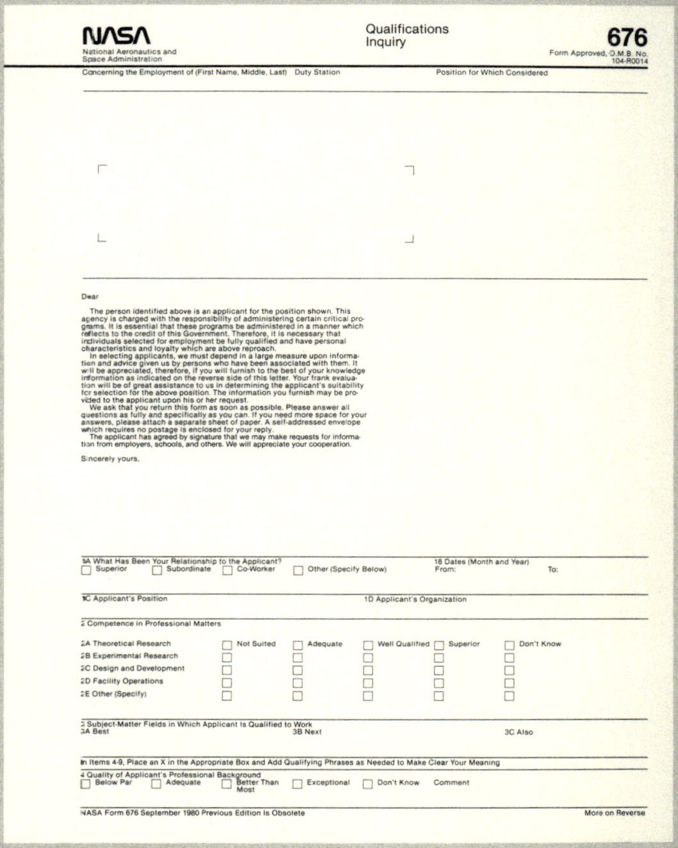

NASA
National Aeronautics and Space Administration
Qualifications Inquiry
676
Form Approved, O.M.B. No. 104-R0014

Concerning the Employment of (First Name, Middle, Last) | Duty Station | Position for Which Considered

Dear

The person identified above is an applicant for the position shown. This agency is charged with the responsibility of administering certain critical programs. It is essential that these programs be administered in a manner which reflects to the credit of this Government. Therefore, it is necessary that individuals selected for employment be fully qualified and have personal characteristics and loyalty which are above reproach.

In selecting applicants, we must depend in a large measure upon information and advice given us by persons who have been associated with them. It will be appreciated, therefore, if you will furnish to the best of your knowledge information as indicated on the reverse side of this letter. Your frank evaluation will be of great assistance to us in determining the applicant's suitability for selection for the above position. The information you furnish may be provided to the applicant upon his or her request.

We ask that you return this form as soon as possible. Please answer all questions as fully and specifically as you can. If you need more space for your answers, please attach a separate sheet of paper. A self-addressed envelope which requires no postage is enclosed for your reply.

The applicant has agreed by signature that we may make requests for information from employers, schools, and others. We will appreciate your cooperation.

Sincerely yours,

1A What Has Been Your Relationship to the Applicant? ☐ Superior ☐ Subordinate ☐ Co-Worker ☐ Other (Specify Below) | 1B Dates (Month and Year) From: To:
1C Applicant's Position | 1D Applicant's Organization
2 Competence in Professional Matters

	Not Suited	Adequate	Well Qualified	Superior	Don't Know
2A Theoretical Research	☐	☐	☐	☐	☐
2B Experimental Research	☐	☐	☐	☐	☐
2C Design and Development	☐	☐	☐	☐	☐
2D Facility Operations	☐	☐	☐	☐	☐
2E Other (Specify)	☐	☐	☐	☐	☐

3 Subject-Matter Fields in Which Applicant Is Qualified to Work 3A Best | 3B Next | 3C Also
In Items 4-9, Place an X in the Appropriate Box and Add Qualifying Phrases as Needed to Make Clear Your Meaning
4 Quality of Applicant's Professional Background ☐ Below Par ☐ Adequate ☐ Better Than Most ☐ Exceptional ☐ Don't Know | Comment
NASA Form 676 September 1980 Previous Edition Is Obsolete | More on Reverse

—Whenever the form requires respondents name and address and subsequent insertion into a window envelope, indicate position with brackets equal in size to two sides of an open box.
—"Letter" text may set as 1 column on a 2 column format.

NASA Business Forms

Because of the density and diversity of NASA forms, certain fundamental guidelines have been developed to assist those persons responsible for forms. For purposes of demonstration, several typical forms have been redesigned and are shown on these pages. These models address many of the primary problems which forms designers face, while still adhering to the basic principles of the NASA Graphic Standards program.

Typography

The typeface Helvetica, in regular and medium weights has been chosen for optimum legibility as well as to conform to the overall NASA Graphic Standards Program. 7/7 pt. Helvetica Regular should be used for the main body of the form and 7/7 pt. Helvetica Medium for emphasis. All typography should flush left with the grid. Instructions, or explanatory text should be set in 7/7 pt. Helvetica Regular initial caps & lower case, flush left (on the grid) and rag right, with 1 em paragraph indents.

Grid

The basis of the NASA forms system is the underlying grid on an 8½" x 11" form. By organizing information on a grid, a form will appear more orderly and less visually confusing, the result of which is efficiency in both the completion process and subsequent information retrieval. The use of a common grid can also mean time and cost savings in forms planning, design, typesetting and production.

The NASA forms grid is based on both the horizontal character spacing and vertical line spacing of a pica (10 pitch) typewriter. In order to take advantage of modern computer typography (which has a maximum typesetting width of 45 picas), the "live area" of the grid is made up of 8 horizontal increments each of which can contain a maximum of 9 typewritten characters. On an engineering scale, this measure represents 9 units on the 10 units per inch scale. Each horizontal increment can also be broken into ½ increments containing a maximum of 4 characters. The vertical increments are based on a pica scale which is equal to 6 units per inch. Thus the live area of the form contains 72 horizontal units and 60 vertical units. A standard 2 pica margin is maintained on the top, bottom and right side of the form. If there is an instance that an odd-sized form is unavoidable, it is suggested that the basic theories of this grid be related to the new size.

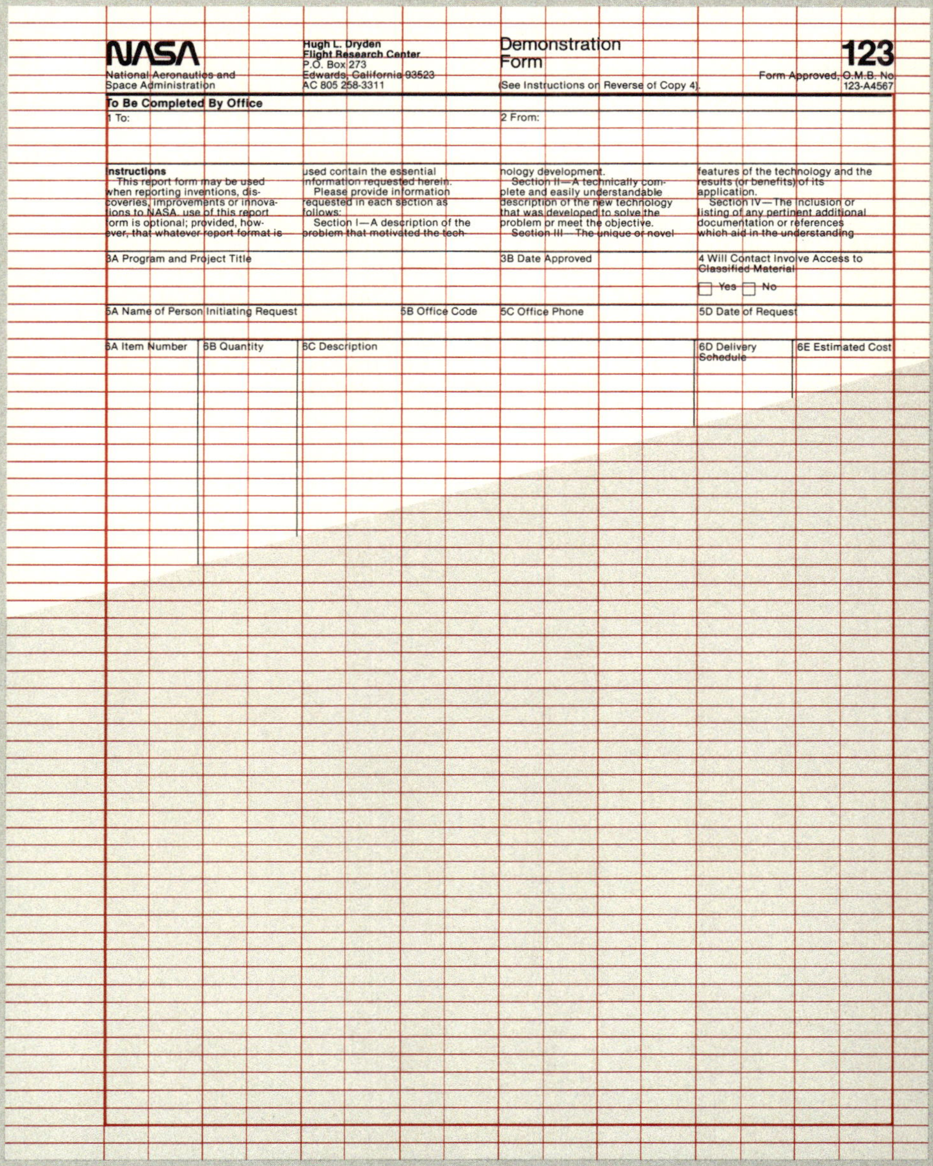
NASA
National Aeronautics and Space Administration

Hugh L. Dryden Flight Research Center
P.O. Box 273
Edwards, California 93523
AC 805 258-3311

Demonstration Form
(See Instructions on Reverse of Copy 4).

123
Form Approved, O.M.B. No. 123-A4567

To Be Completed By Office

1 To:

2 From:

Instructions
This report form may be used when reporting inventions, discoveries, improvements or innovations to NASA, use of this report form is optional; provided, however, that whatever report format is used contain the essential information requested herein.
Please provide information requested in each section as follows:
Section I—A description of the problem that motivated the technology development.
Section II—A technically complete and easily understandable description of the new technology that was developed to solve the problem or meet the objective.
Section III—The unique or novel features of the technology and the results (or benefits) of its application.
Section IV—The inclusion or listing of any pertinent additional documentation or references which aid in the understanding

3A Program and Project Title

3B Date Approved

4 Will Contact Involve Access to Classified Material
☐ Yes ☐ No

5A Name of Person Initiating Request

5B Office Code

5C Office Phone

5D Date of Request

6A Item Number	6B Quantity	6C Description	6D Delivery Schedule	6E Estimated Cost

—Masthead may only contain the information indicated.
—Position vertical rules 1/20 of an inch to the left of typography.
—Instructional text sets to a 4-column format, align bottom lines whenever possible.

typewriter carriage movement. Whenever possible, related information should be arranged together on the same line in a logical order (for example: name, signature, date). Careful consideration should be given to the number of lines necessary to complete an answer.

Language
Because of the complex nature of forms, language should be as clear and concise as possible. Likewise, minimizing the amount of writing (or typing) to complete the form will aid in making the completion process more efficient.

A numerical ordering system is advisable to: establish a clear secuence of response; provide an efficient method of reference and retrieval; provide another means of relating groups of information. Whenever there are two or more pieces of related information, the individual items should be designated with both a number and an alphabetical prefix. In instances where related information is grouped under a common head, the head should be designated by a number and the items should receive the same number and a prefix. In order to avoid confusion, all items requiring a response should be designated as suggested. The only exception is that boxes should not be numbered, as they are essentially a response to a question, and not a question in itself.

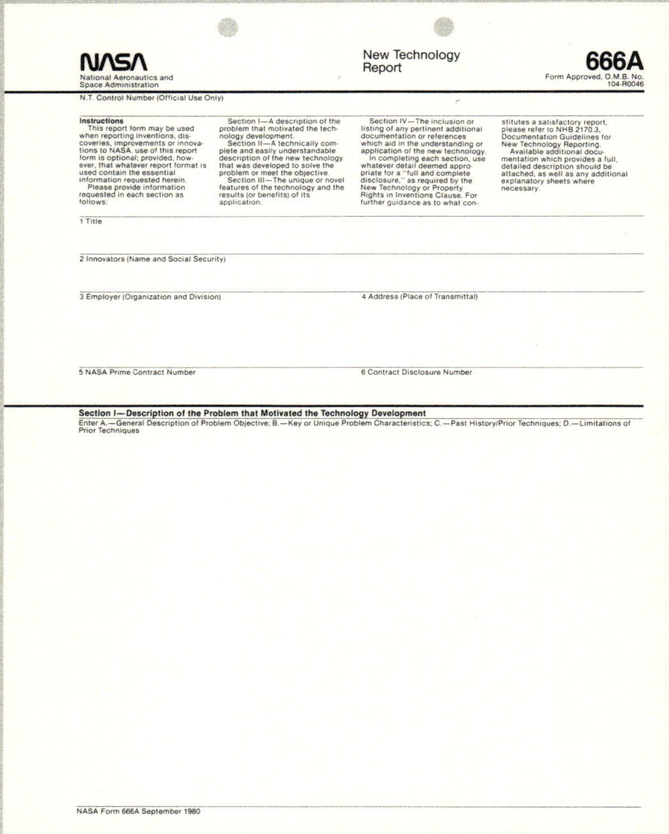
NASA
National Aeronautics and Space Administration

New Technology Report

666A
Form Approved, O.M.B. No. 104-R0046

N.T. Control Number (Official Use Only)

Instructions
This report form may be used when reporting inventions, discoveries, improvements or innovations to NASA. use of this report form is optional; provided, however, that whatever report format is used contain the essential information requested herein.
Please provide information requested in each section as follows:
Section I—A description of the problem that motivated the technology development.
Section II—A technically complete and easily understandable description of the new technology that was developed to solve the problem or meet the objective.
Section III—The unique or novel features of the technology and the results (or benefits) of its application.
Section IV—The inclusion or listing of any pertinent additional documentation or references which aid in the understanding or application of the new technology.
In completing each section, use whatever detail deemed appropriate for a "full and complete disclosure," as required by the New Technology or Property Rights in Inventions Clause. For further guidance as to what constitutes a satisfactory report, please refer to NHB 2170.3, Documentation Guidelines for New Technology Reporting.
Available additional documentation which provides a full, detailed description should be attached, as well as any additional explanatory sheets where necessary.

1 Title

2 Innovators (Name and Social Security)

3 Employer (Organization and Division) | 4 Address (Place of Transmittal)

5 NASA Prime Contract Number | 6 Contract Disclosure Number

Section I—Description of the Problem that Motivated the Technology Development
Enter A.—General Description of Problem Objective; B.—Key or Unique Problem Characteristics; C.—Past History/Prior Techniques; D.—Limitations of Prior Techniques

NASA Form 666A September 1980

—Shift format down 2 increments to allow for a 2 hole punch at the top.
—Text sets 10 pica maximum.
—When instructional text is short (see Section I) it may set to the full width for maximum readability.
—Leave a sufficiently large space (as in Section I) to allow for a lengthy response.

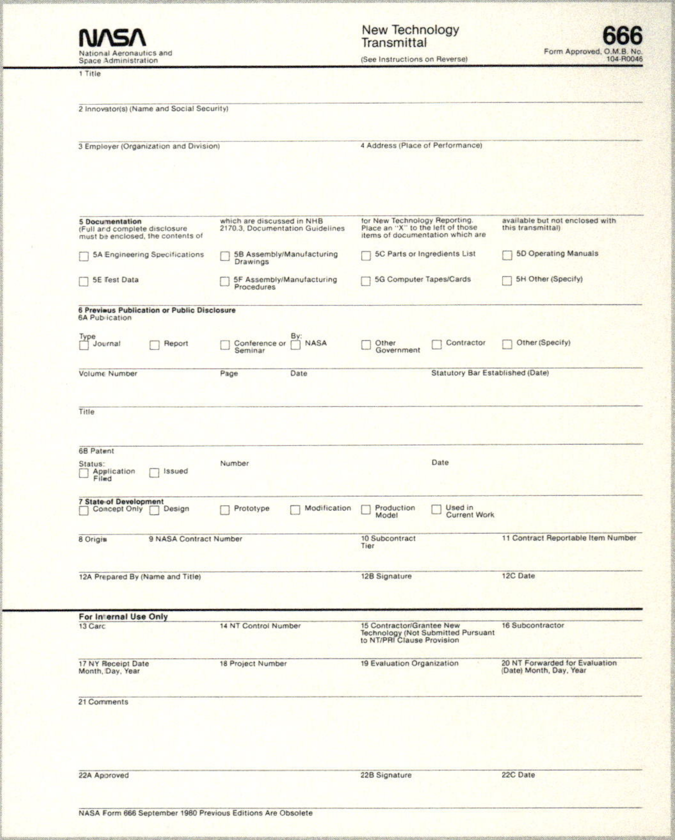
NASA
National Aeronautics and Space Administration

New Technology Transmittal
(See Instructions on Reverse)

666
Form Approved, O.M.B. No. 104-R0046

1 Title

2 Innovator(s) (Name and Social Security)

3 Employer (Organization and Division) | 4 Address (Place of Performance)

5 Documentation
(Full and complete disclosure must be enclosed, the contents of which are discussed in NHB 2170.3, Documentation Guidelines for New Technology Reporting. Place an "X" to the left of those items of documentation which are available but not enclosed with this transmittal)

☐ 5A Engineering Specifications ☐ 5B Assembly/Manufacturing Drawings ☐ 5C Parts or Ingredients List ☐ 5D Operating Manuals
☐ 5E Test Data ☐ 5F Assembly/Manufacturing Procedures ☐ 5G Computer Tapes/Cards ☐ 5H Other (Specify)

6 Previous Publication or Public Disclosure
6A Publication
Type: ☐ Journal ☐ Report ☐ Conference or Seminar
By: ☐ NASA ☐ Other Government ☐ Contractor ☐ Other (Specify)

Volume Number | Page | Date | Statutory Bar Established (Date)

Title

6B Patent
Status: ☐ Application Filed ☐ Issued | Number | Date

7 State of Development
☐ Concept Only ☐ Design ☐ Prototype ☐ Modification ☐ Production Model ☐ Used in Current Work

8 Origin | 9 NASA Contract Number | 10 Subcontract Tier | 11 Contract Reportable Item Number

12A Prepared By (Name and Title) | 12B Signature | 12C Date

For Internal Use Only
13 Card | 14 NT Control Number | 15 Contractor/Grantee New Technology (Not Submitted Pursuant to NT/PRI Clause Provision | 16 Subcontractor

17 NY Receipt Date Month, Day, Year | 18 Project Number | 19 Evaluation Organization | 20 NT Forwarded for Evaluation (Date) Month, Day, Year

21 Comments

22A Approved | 22B Signature | 22C Date

NASA Form 666 September 1980 Previous Editions Are Obsolete

—Emphasize specific items with 7 pt. Helvetica Medium.
—Organize material in left to right configuration, wherever possible, for ease of entries.

alternatives. Whenever possible, boxes should be arranged horizontally, both to avoid vertical carriage movement and unnecessary use of vertical space.

Masthead

The major purpose of the masthead is form identification. Strict adherence to these standards is required. The masthead consists of three primary elements and three secondary elements. The primary elements are: a 16 pt. cap height NASA logotype with a 7/7 pt. Helvetica Regular signature, a 12/12 pt. Helvetica Regular form title and 24 pt. Helvetica Medium form number. The secondary elements are as follows: Upon approval from NASA Headquarters, NASA Center designations may appear, setting in 7/7 pt. Helvetica Medium with the address and phone number in 7/7 pt. Helvetica Regular. When necessary, "See instructions on Reverse" and the OMB Approval line set 7/7 pt. Helvetica Regular.

Bottom Margin

The bottom margin should contain a maximum of three specific pieces of information. The NASA form number and revision date should appear flush left on the left side in 7/7 pt. Helvetica Regular.

Layout

Organize material in a left to right and top to bottom sequence to minimize

NASA 843
National Aeronautics and Space Administration
For Internal Routing Only
Code | Name | Date | Remarks

Confidential

The attachment contains information affecting the National Defense of the United States within the meaning of the espionage laws, Title 18, U.S.C., Sections 793 and 794, the transmission or revelation of which in any manner to an unauthorized person is prohibited by law.
This form must be attached to all Confidential documents not stored in security storage equipment.
Unclassified when detached from Classified Material.

NASA Form 843 (Revised September 1980)

—Allow 3 vertical increments between rules for handwritten forms.
—Text type is 7/7 pt. Helvetica Medium for emphasis.
—Form should print red on yellow paper stock to distinguish it from other material.

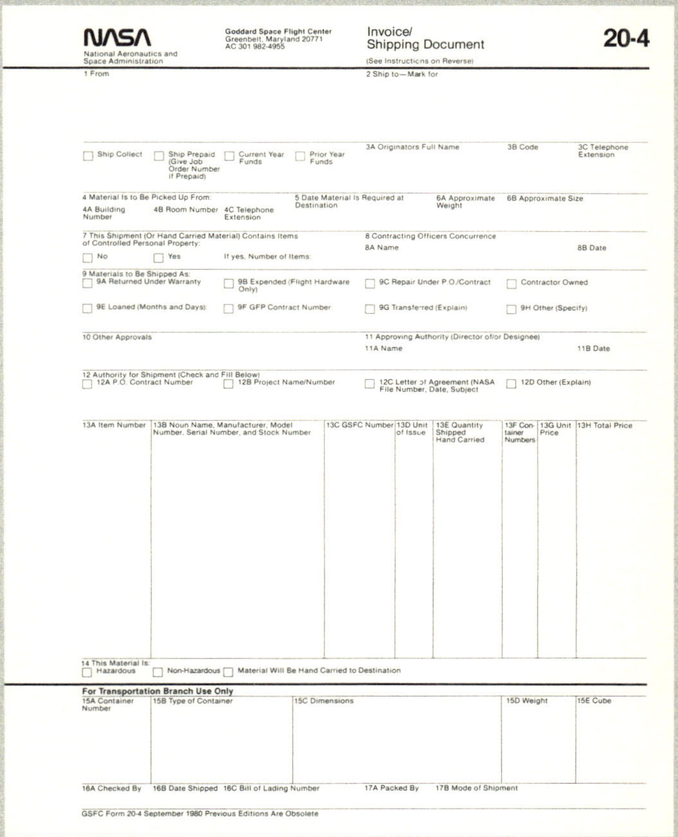

NASA 20-4
National Aeronautics and Space Administration
Goddard Space Flight Center
Greenbelt, Maryland 20771
AC 301 982-4955
Invoice/
Shipping Document
(See Instructions on Reverse)

1 From
2 Ship to—Mark for
Ship Collect | Ship Prepaid (Give Job Order Number if Prepaid) | Current Year Funds | Prior Year Funds
3A Originators Full Name | 3B Code | 3C Telephone Extension
4 Material Is to Be Picked Up From:
4A Building Number | 4B Room Number | 4C Telephone Extension
5 Date Material Is Required at Destination
6A Approximate Weight | 6B Approximate Size
7 This Shipment (Or Hand Carried Material) Contains Items of Controlled Personal Property:
No | Yes | If yes, Number of Items:
8 Contracting Officers Concurrence
8A Name | 8B Date
9 Materials to Be Shipped As:
9A Returned Under Warranty | 9B Expended (Flight Hardware Only) | 9C Repair Under P.O./Contract | Contractor Owned
9E Loaned (Months and Days): | 9F GFP Contract Number: | 9G Transferred (Explain) | 9H Other (Specify)
10 Other Approvals
11 Approving Authority (Director of/or Designee)
11A Name | 11B Date
12 Authority for Shipment (Check and Fill Below)
12A P.O. Contract Number | 12B Project Name/Number | 12C Letter of Agreement (NASA File Number, Date, Subject | 12D Other (Explain)

13A Item Number	13B Noun Name, Manufacturer, Model Number, Serial Number, and Stock Number	13C GSFC Number	13D Unit of Issue	13E Quantity Shipped Hand Carried	13F Container Numbers	13G Unit Price	13H Total Price

14 This Material Is:
Hazardous | Non-Hazardous | Material Will Be Hand Carried to Destination

For Transportation Branch Use Only

15A Container Number	15B Type of Container	15C Dimensions	15D Weight	15E Cube

16A Checked By | 16B Date Shipped | 16C Bill of Lading Number | 17A Packed By | 17B Mode of Shipment

GSFC Form 20-4 September 1980 Previous Editions Are Obsolete

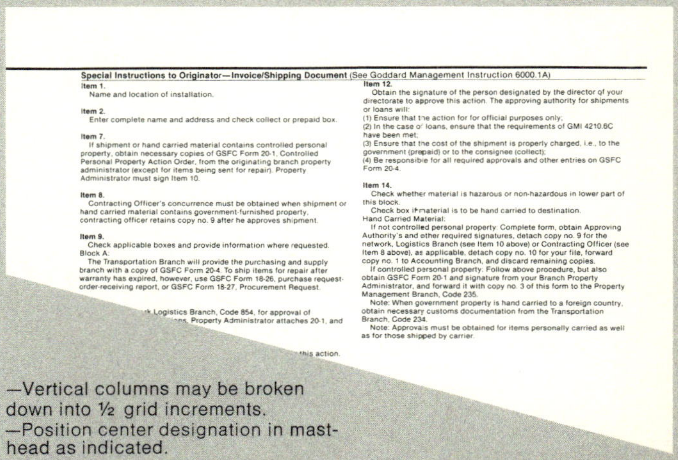

Special Instructions to Originator—Invoice/Shipping Document (See Goddard Management Instruction 6000.1A)

Item 1.
Name and location of installation.

Item 2.
Enter complete name and address and check collect or prepaid box.

Item 7.
If shipment or hand carried material contains controlled personal property, obtain necessary copies of GSFC Form 20-1, Controlled Personal Property Action Order, from the originating branch property administrator (except for items being sent for repair). Property Administrator must sign Item 10.

Item 8.
Contracting Officer's concurrence must be obtained when shipment or hand carried material contains government-furnished property, contracting officer retains copy no. 9 after he approves shipment.

Item 9.
Check applicable boxes and provide information where requested.
Block A:
The Transportation Branch will provide the purchasing and supply branch with a copy of GSFC Form 20-4. To ship items for repair after warranty has expired, however, use GSFC Form 18-26, purchase request-order-receiving report, or GSFC Form 18-27, Procurement Request.

...k Logistics Branch, Code 854, for approval of ...ons, Property Administrator attaches 20-1, and

...this action.

Item 12.
Obtain the signature of the person designated by the director of your directorate to approve this action. The approving authority for shipments or loans will:
(1) Ensure that the action for for official purposes only;
(2) In the case of loans, ensure that the requirements of GMI 4210.6C have been met;
(3) Ensure that the cost of the shipment is properly charged, i.e., to the government (prepaid) or to the consignee (collect);
(4) Be responsible for all required approvals and other entries on GSFC Form 20-4.

Item 14.
Check whether material is hazarous or non-hazardous in lower part of this block.
Check box if material is to be hand carried to destination.
Hand Carried Material:
If not controlled personal property: Complete form, obtain Approving Authority's and other required signatures, detach copy no. 9 for the network, Logistics Branch (see Item 10 above) or Contracting Officer (see Item 8 above), as applicable, detach copy no. 10 for your file, forward copy no. 1 to Accounting Branch, and discard remaining copies.
If controlled personal property: Follow above procedure, but also obtain GSFC Form 20-1 and signature from your Branch Property Administrator, and forward it with copy no. 3 of this form to the Property Management Branch, Code 235.
Note: When government property is hand carried to a foreign country, obtain necessary customs documentation from the Transportation Branch, Code 234.
Note: Approvals must be obtained for items personally carried as well as for those shipped by carrier.

—Vertical columns may be broken down into ½ grid increments.
—Position center designation in masthead as indicated.
—Instructional text on the backside should set in 2 column format at 20½ pica maximum.
—Backside should align head to head with front side.

5

NASA Publications

Publications constitute the largest communications output from NASA. It is in the best interests of NASA to maintain a visual consistency and high level of design excellence in all of its publications, whether they are large or small in their scope and in terms of the audiences they are intended to reach. A consistency must also exist in the way that NASA and NASA Centers identify themselves as originators of publications.

In this way, each publication itself becomes an important extension of NASA's identity effort.

The guidelines set forth in this section are intended to be an aid in accomplishing a strong, integrated, consistent "family" of publications throughout NASA.

Because of the number and diverse nature of NASA's publications, the guidelines are general for the most part. They are, however, specific with respect to certain elements of style, use of typography and use of the NASA logotype.

The NASA Logotype

The NASA logotype and its attendant elements of style and identification are always a required element in any NASA publication. Somewhere in the publication full NASA identification must be shown in order to clarify the origin of the publication and to identify NASA.

This identification does not necessarily have to appear as part of a front cover design, although in many instances this approach seems appropriate. It may appear on the last page of a publication, or on the back cover. The goal is to insure that each publication issued by the agency is clearly identified and credited to NASA.

NASA "Stem-Word"

In some examples shown in this section, the NASA logotype has been used as a "stem-word" (NASA Activities, NASA Facts, etc.).

This allows the logotype to become at once the major identification element and part of the publication title.

The stem-word technique is used only on permanent, regularly published documents, and requires advice from the NASA Graphics Coordinator at Headquarters.

Illustration & Photography

Use of illustration & photography as either story telling or major graphic elements is a key in the design of NASA publications.

Simplicity, appropriateness and strength of composition is important in the successful handling of these elements, not technique alone as is sometimes believed and practiced.

The absence of detail and deliniation, or symbolic quality, in a cover illustration or photograph provokes interest and stimulates the reader to go to the inside of the publication for more information.

In the case of a technical diagram or story-telling illustration, correct detailing is absolutely necessary to accomplish the desired communication.

Always analyze the illustration in terms of the communication task it must accomplish. Then determine what technique to use, how realistic it must be, etc. Where possible, within the restrictions of budget and available material, keep your options open and utilize the best available photography and illustration talent. Make certain they are fully briefed on the task you are assigning to them and that they understand the broad communication goal of the project before beginning their work.

Several examples of different sorts of illustration and photography are shown in the following pages of this section. Note the relationship of illustration to typography and how it is used in relation to the total image produced, not only how successful it is in and of itself.

Typography

Typography and typographic design are the "architecture" of any publication. Based on the structure of typography, the various elements in a publication fall into their logical locations and relationships, forming in the end a harmonious sequence of visual events.

In this section are four pages devoted to recommended NASA type styles and sample settings of each. Each

style has a specific reason for inclusion in the vocabulary and should be used accordingly.

The most important specific typographic style is derived from the elements of NASA identification and is composed of Helvetica Light and Helvetica Medium, upper and lower case. This should be the most used style in NASA publications because of its direct relationship to the NASA identification elements.

Recommended general elements of typographic style include flush left, ragged right column setting, bold face headlines and subheadlines and minimum one-point line spacing in text setting.

Color

Government printing regulations relegate most NASA publications to a one or two color printing limitation. The exception to this rule occurs in certain special situations where the use of 4/color process printing is appropriate, desirable or necessary for the effective portrayal of a particular story.

When limited to one color printing, middle range to dark colors should be used to cover the normal range of reproduction and legibility requirements, including use of the NASA logotype.

When a second color is available for use with black, three options are presented. First, NASA red may be used to accentuate the NASA identification and to perform as an accent color in the publication design. Second, one of a range of bright, primary and secondary colors may be used to enhance the design or aid in the presentation of information. The use of pale, pastel colors is discouraged in all NASA publications. Third, NASA gray or another middle value gray or neutral color may be used as a second color for manufacturing "duo tones" or two-color half-tone renderings. In situations where black and white photography dominates, this technique tends to strengthen and deepen the photographic images, making them much more aesthetically pleasing than black half-tones.

In some special situations, 4/color process printing is available. NASA identification may be easily portrayed in color through the use of 4/color ben-day mixtures.

Review the information on color in section one of this manual.

Paper

Within the limitations of budget, availability and government printing standards, paper specifications for NASA publications should be directed toward high-quality non-absorbent (coated or non-coated) papers which provide good ink hold-out and do not produce fuzzy images as a result of their surface quality. Colors specified should not be in the pastel or faded color range, but rather bright primary and secondary colors and in the middle value range.

Sizes

Sizes of NASA publications should, as a general rule, be determined on the basis of standard government printing sizes and capabilities.

Deviations from standard sizes should only be used in instances where the subject matter or the occasion demand a special treatment.

Most of the publications shown in this section are based on standard government printing sizes.

Formats

All of NASA's publication requirements, especially in the booklet and brochure category, should utilize a vertical format unless the subject matter or occasion provides a compelling reason to deviate. Horizontal formats are normally clumsy to handle, more awkward to design and there is no appreciable gain in the total effect produced by using such a format.

Consult with the NASA Graphics Coordinator at Headquarters on questions of size and formats for NASA publications as well as any question generally or specifically related to these guidelines.

Typography—Sans Serif
Helvetica

Helvetica is the most important family of type in the NASA Unified Visual Communications System. Helvetica Light is used in combination with the logotype to form the fundamental elements of identification.

In addition, this typeface can be used in numerous media and in a variety of situations to create a clean and contemporary visual program. The cursive san-serif letterforms make it extremely legible, even at very small sizes.

Headings which accompany Helvetica Light text settings are set in Helvetica Medium. In certain situations Helvetica Bold may be an appropriate alternative. Headings are set in upper and lower case.

ABCDEFGHIJKLMNOPQRSTUVWXYZ
abcdefghijklmnopqrstuvwxyz
1234567890$(&?!%.,:;-)

Helvetica Light

ABCDEFGHIJKLMNOPQRSTUVWXYZ
abcdefghijklmnopqrstuvwxyz
1234567890$(&?!%.,:;-)

Helvetica Medium

---

10 pt. Helvetica Light; 10, 14 pt. Helvetica Medium

Helvetica Medium

The main purpose of letters is the practical one of making thoughts visible. Ruskin says that all letters are frightful things, and to be endured only upon occasion, that is to say, in places where the sense of the inscription is of more importance than external ornament. This is a sweeping statement from which we need not suffer unduly; yet it is doubtful if there is art in individual letters. Letters in combination may be quite satisfying and in a well-composed page beautiful as a whole. The main purpose of letters is the practical one of making thoughts visible. Ruskin says that all letters are frightful things, and to be endured only upon occasion, that is to say, in places where the sense of the

Helvetica Medium

Ruskin says that all letters are frightful things, and to be endured only upon occasion, that is to say, in places where the sense of the inscription is of more importance than external ornament. This is a sweeping statement from which we need not suffer unduly; yet it is doubtful if there is art in individual letters. Letters in combination may be quite satisfying and in a well-composed page beautiful as a whole. The main purpose of letters is the practical one of making

10 pt. Helvetica Light; 10, 14 pt. Helvetica Medium

Helvetica Medium

The main purpose of letters is the practical one of making thoughts visible. Ruskin says that all letters are frightful things, and to be endured only upon occasion, that is to say, in places where the sense of the inscription is of more importance than external ornament. This is a sweeping statement from which we need not

Helvetica Medium

Ruskin says that all letters are frightful things, and to be endured only upon occasion, that is to say, in places where the sense of the inscription is of more importance than external ornament. This is a sweeping statement from which we need not suffer unduly; yet it is doubtful if there is art in individual letters. Letters in combination may be quite satisfying and in a well-composed page

Futura is recommended for a number of reasons. The typeface is quite legible and is versatile enough for catalog listings as well as brochure applications. The precision letterforms have a technological character and make it a natural for certain NASA projects.

The Futura face is designed with a small x-height and will require special attention when specifying the size. In the comparison of typefaces enclosed in this section, you will note that 11 pt. Futura is comparable in appearance to 10 pt. Helvetica.

When the Futura face is being used, always specify Futura Demibold headings. Do not mix Helvetica Medium headings with Futura text settings.

ABCDEFGHIJKLMNOPQRSTUVWXYZ
abcdefghijklmnopqrstuvwxyz
1234567890$(&?!%.,:;-)

Futura Light

ABCDEFGHIJKLMNOPQRSTUVWXYZ
abcdefghijklmnopqrstuvwxyz
1234567890$(&?!%.,:;-)

Futura Demibold

---

11 pt. Futura Light; 11, 14 pt. Futura Demibold

Futura Demibold
The main purpose of letters is the practical one of making thoughts visible. Ruskin says that all letters are frightful things, and to be endured only upon occasion, that is to say, in places where the sense of the inscription is of more importance than external ornament. This is a sweeping statement from which we need not suffer unduly; yet it is doubtful if there is art in individual letters. Letters in combination may be quite satisfying and in a well-composed page beautiful as a whole. The main purpose of letters is the practical one of making thoughts visible. Ruskin says that all letters are frightful things, and to be endured only upon occasion, that is to say, in places where the sense of the inscription is of more importance than external ornament.
Futura Demibold
Ruskin says that all letters are frightful things, and to be endured only upon occasion, that is to say, in places where the sense of the inscription is of more importance than external ornament. This is a sweeping statement from which we need not suffer unduly; yet it is doubtful if there is art in individual letters. Letters in combination may be quite satisfying and in a well-composed page beautiful as a whole. The main purpose of letters is the practical one of making thoughts visible. Ruskin

11 pt. Futura Light; 11, 14 pt. Futura Demibold

Futura Demibold
The main purpose of letters is the practical one of making thoughts visible. Ruskin says that all letters are frightful things, and to be endured only upon occasion, that is to say, in places where the sense of the inscription is of more importance than external ornament. This is a sweeping statement from which we need not suffer unduly; yet it is doubtful
Futura Demibold
Ruskin says that all letters are frightful things, and to be endured only upon occasion, that is to say, in places where the sense of the inscription is of more importance than external ornament. This is a sweeping statement from which we need not suffer unduly; yet it is doubtful if there is art in individual letters. Letters in combination may be quite satisfying and in a well-composed page beautiful as a whole. The main purpose of letters is the

Typography—Serif
Garamond

Garamond is perhaps the finest of the "classical" typefaces. It has stood the test of time and proved itself to be as useful in contemporary design as it has been in more traditional applications.

The main virtues of Garamond include superior readability, handsome character, a distinctive Italic, and certain special refinements such as old style numerals.

Garamond is ideal for high quality publications or those of a more permanent nature. It functions very well in large volume settings and will sustain reader attention.

Headings may be set in Helvetica Medium or Garamond Bold. Garamond Bold is not a particularly heavy weight and the designer must compensate by increasing the size of the headings or by using space around them.

ABCDEFGHIJKLMNOPQRSTUVWXYZ
abcdefghijklmnopqrstuvwxyz
1234567890$(&?!%.,:;-)

Garamond

ABCDEFGHIJKLMNOPQRSTUVWXYZ
abcdefghijklmnopqrstuvwxyz
1234567890$(&?!%.,:;-)

Garamond Bold

---

10 pt. Garamond; 10, 14 pt. Helvetica Medium

Helvetica Medium

The main purpose of letters is the practical one of making thoughts visible. Ruskin says that all letters are frightful things, and to be endured only upon occasion, that is to say, in places where the sense of the inscription is of more importance than external ornament. This is a sweeping statement from which we need not suffer unduly; yet it is doubtful if there is art in individual letters. Letters in combination may be quite satisfying and in a well-composed page beautiful as a whole. The main purpose of letters is the practical one of making thoughts visible. Ruskin says that all letters are frightful things, and to be endured only upon occasion, that is to say, in places where the sense of the inscription is of more importance than external ornament. This is a sweeping statement from which we

Helvetica Medium

Ruskin says that all letters are frightful things, and to be endured only upon occasion, that is to say, in places where the sense of the inscription is of more importance than external ornament. This is a sweeping statement from which we need not suffer unduly; yet it is doubtful if there is art in individual letters. Letters in combination may be quite satisfying and in a well-composed page beautiful as a whole. The main purpose of letters is the practical one of making thoughts visible. Ruskin says that all letters are

10 pt. Garamond; 11, 14 pt. Garamond Bold

Garamond Bold

The main purpose of letters is the practical one of making thoughts visible. Ruskin says that all letters are frightful things, and to be endured only upon occasion, that is to say, in places where the sense of the inscription is of more importance than external ornament. This is a sweeping statement from which we need not suffer unduly; yet it is doubtful if there is art in

Garamond Bold

Ruskin says that all letters are frightful things, and to be endured only upon occasion, that is to say, in places where the sense of the inscription is of more importance than external ornament. This is a sweeping statement from which we need not suffer unduly; yet it is doubtful if there is art in individual letters. Letters in combination may be quite satisfying and in a well-composed page beautiful as a whole. The main purpose of letters is the practical

Typography—Serif
Times Roman

Times Roman is generally regarded as the best of the modern or transitional typefaces. It offers readability, character, and a certain utilitarian quality which makes it quite useful in publication design. It is designed with a large x-height which makes it legible at small sizes.

Times Roman is recommended for newsletters, house organs and other news-oriented publications. This typeface is appropriate for large volume settings as the reader does not tire of the appearance.

The entire family of Times Roman, including Italic and Bold, gives the designer a practical typeface to solve certain complicated problems.

Headings may be set in Helvetica Medium or in Times Roman Bold.

ABCDEFGHIJKLMNOPQRSTUVWXYZ
abcdefghijklmnopqrstuvwxyz
1234567890$(&?!%.,:;-)

Times Roman

ABCDEFGHIJKLMNOPQRSTUVWXYZ
abcdefghijklmnopqrstuvwxyz
1234567890$(&?!%.,:;-)

Times Roman Bold

---

10 pt. Times Roman; 10, 14 pt. Helvetica Medium

Helvetica Medium

The main purpose of letters is the practical one of making thoughts visible. Ruskin says that all letters are frightful things, and to be endured only upon occasion, that is to say, in places where the sense of the inscription is of more importance than external ornament. This is a sweeping statement from which we need not suffer unduly; yet it is doubtful if there is art in individual letters. Letters in combination may be quite satisfying and in a well-composed page beautiful as a whole. The main purpose of letters is the practical one of making thoughts visible. Ruskin says that all letters are frightful things, and to be endured only upon occasion, that is to say, in places where the sense of the inscription is of more importance than external

Helvetica Medium

Ruskin says that all letters are frightful things, and to be endured only upon occasion, that is to say, in places where the sense of the inscription is of more importance than external ornament. This is a sweeping statement from which we need not suffer unduly; yet it is doubtful if there is art in individual letters. Letters in combination may be quite satisfying and in a well-composed page beautiful as a whole. The main purpose of letters is the practical one of making thoughts visible.

10 pt. Times Roman; 10, 14 pt. Times Roman Bold

Times Roman Bold

The main purpose of letters is the practical one of making thoughts visible. Ruskin says that all letters are frightful things, and to be endured only upon occasion, that is to say, in places where the sense of the inscription is of more importance than external ornament. This is a sweeping statement from which we need not suffer unduly; yet it

Times Roman Bold

Ruskin says that all letters are frightful things, and to be endured only upon occasion, that is to say, in places where the sense of the inscription is of more importance than external ornament. This is a sweeping statement from which we need not suffer unduly; yet it is doubtful if there is art in individual letters. Letters in combination may be quite satisfying and in a well-composed page beautiful as a whole. The main

Cover Design: Leaflets & Folders

Covers for small-scale publications should be thought of as posters in miniature. Because of the small amount of space, the designer should attempt to reduce the number of competitive elements and to strive for simplicity. This will require the cooperation of editorial colleagues so that the cover can be clear, direct and attractive.

When brochures are to be published in a series, it is advisable to plan ahead so that a coordinated family look might emerge.

In the demonstrations below, note that all examples incorporate the NASA logotype and identification but maintain their basic simplicity. The proper coordination of type, photography and illustration will ensure that small covers have the impact of larger pieces.

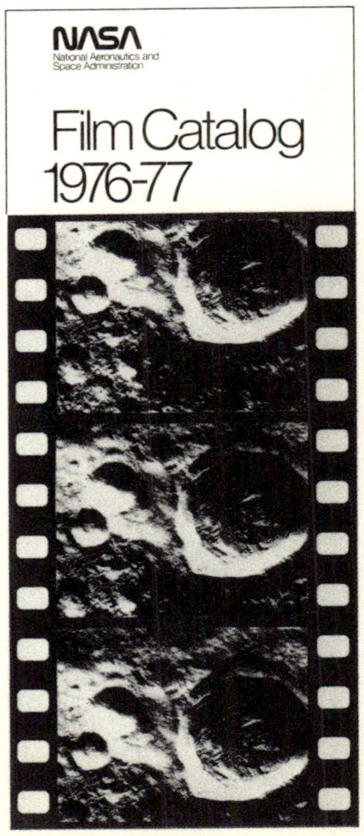

Cover Design: Journals and Technical Publications

Covers for publications in this category are straightforward, simple and devoid of frills. This is the ideal approach to publications which are vehicles for scientific information and research data.

The covers shown below employ the stem-word use of the NASA logotype. Competitive elements have been reduced and attention has been given to proper emphasis on Center accreditation and headline treatment. Standards, such as the use of an outline box around all diagrams, will help organize the covers. Catalog numbers should always appear in the same location from issue to issue.

NASA Tech Briefs
National Aeronautics and Space Administration
Ut enim ad minim veniam, quis laboris nisi ut aliquip ex ea irure dolor in reprehenderit in v illum dolore eu fugiat nulla dignissum qui blandit est praes molestias excepteur sint oc sunt in culpa qui officia deserur Et harumd dereud facilis est

Front cover

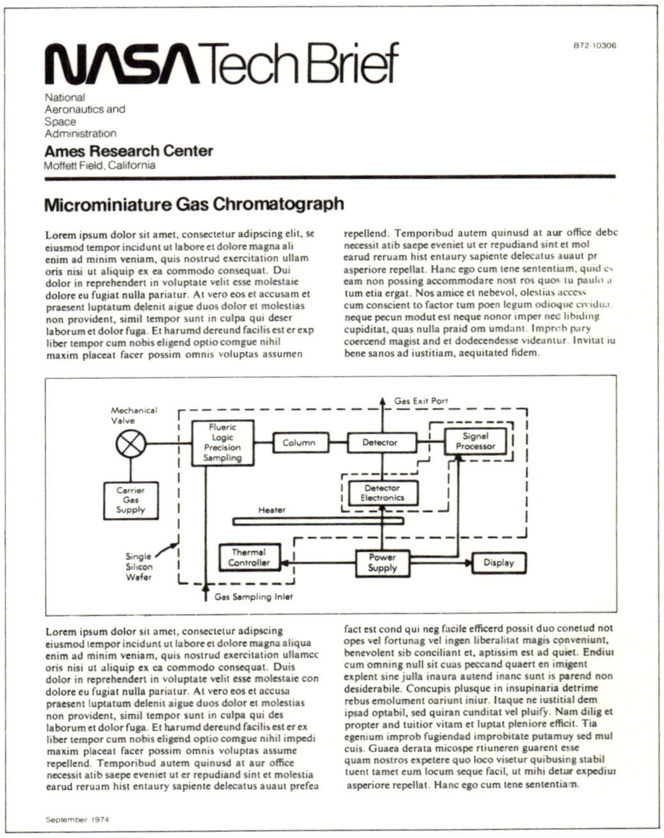
NASA Tech Brief

B72-10306

National Aeronautics and Space Administration

Ames Research Center
Moffett Field, California

Microminiature Gas Chromatograph

Lorem ipsum dolor sit amet, consectetur adipscing elit, se eiusmod tempor incidunt ut labore et dolore magna ali enim ad minim veniam, quis nostrud exercitation ullam oris nisi ut aliquip ex ea commodo consequat. Dui dolor in reprehendert in voluptate velit esse molestaie dolore eu fugiat nulla pariatur. At vero eos et accusam et praesent luptatum delenit aigue duos dolor et molestias non provident, simil tempor sunt in culpa qui deser laborum et dolor fuga. Et harumd dereund facilis est er exp liber tempor cum nobis eligend optio comgue nihil maxim placeat facer possim omnis voluptas assumen repellend. Temporibud autem quinusd at aur office debc necessit atib saepe eveniet ut er repudiand sint et mol earud reruam hist entaury sapiente delecatus auaut pr asperiore repellat. Hanc ego cum tene sententiam, quid e eam non possing accommodare nost ros quos tu paulo a tum etia ergat. Nos amice et nebevol, olestias access cum conscient to factor tum poen legum odioque cividuı neque pecun modut est neque nonor imper nec libiding cupiditat, quas nulla praid om umdant. Improb pary coercend magist and et dodecendesse videantur. Invitat iu bene sanos ad iustitiam, aequitated fidem.

Lorem ipsum dolor sit amet, consectetur adipscing eiusmod tempor incidunt ut labore et dolore magna aliqua enim ad minim veniam, quis nostrud exercitation ullamcc oris nisi ut aliquip ex ea commodo consequat. Duis dolor in reprehendert in voluptate velit esse molestaie con dolore eu fugiat nulla pariatur. At vero eos et accusa praesent luptatum delenit aigue duos dolor et molestias non provident, simil tempor sunt in culpa qui des laborum et dolor fuga. Et harumd dereund facilis est er ex liber tempor cum nobis eligend optio comgue nihil impedi maxim placeat facer possim omnis voluptas assume repellend. Temporibud autem quinusd at aur office necessit atib saepe eveniet ut er repudiand sint et molestia earud reruam hist entaury sapiente delecatus auaut prefea fact est cond qui neg facile efficerd possit duo conetud not opes vel fortunag vel ingen liberalitat magis conveniunt, benevolent sib conciliant et, aptissim est ad quiet. Endiu cum omning null sit cuas peccand quaert en imigent explent sine julla inaura autend inanc sunt is parend non desiderabile. Concupis plusque in insupinaria detrime rebus emolument oariunt iniur. Itaque ne iustitial dem ipsad optabil, sed quiran cunditat vel pluify. Nam dilig et propter and tuitior vitam et luptat pleniore efficit. Tia egenium improb fugiendad improbitate putamuy sed mul cuis. Guaea derata micospe rtiuneren guarent esse quam nostros expetere quo loco visetur quibusing stabil tuent tamet eum locum seque facil, ut mihi detur expediu asperiore repellat. Hanc ego cum tene sententiam.

September 1974

Title page of supplement

NASA Facts
An Educational Publication of the National Aeronautics and Space Administration

Goddard Space Flight Center
Greenbelt, Maryland

Lorem ipsum eiusmod tempor incidunt ut enim ad minim veniam

Cover Design: News Publications

Mastheads for news publications are strong, attractive and uncluttered. Attention should be paid to the various typographic elements so that things are read in the proper order. Ownership of the publication is paramount and should be followed by the titling of a lead article.

Photographs or illustrations which appear on covers should be selected for their content, impact and graphic value. Unless the subject possesses these attributes, it will be an ineffective cover image.

The covers below demonstrate the use of the NASA logotype on typical Center and agency-wide publications.

National Aeronautics and
Space Administration
Langley Research Center

NASA

Langley Researcher

Volume 14, Issue 2 March 7, 1975

Honeywell Showing Rescheduled

The first propulsive stage and the instrument unit for the Saturn IB launch vehicle which will be used in the Apollo-Soyuz Test Project (ASTP) next summer have been removed from storage at the NASA-Kennedy Space Center (KSC), and post-storage checkout is underway.

The vehicle (designated SA-210), developed under direction of the NASA Marshall Space Flight Center (MSFC), will launch three U.S. astronauts aboard a modified Apollo command module on July 15, 1975, to rendezvous and dock with the Soviet Union's Soyuz spacecraft, which will be launched from the U.S.S.R. seven and one-half hours earlier.

The first stage, designated SIB-210, and the instrument unit, designated IU-210, were taken from inside their environmentally protected cocoon-like shrouds in the Vehicle Assembly building (VAB) where they had been stored since Spring.

Post-storage checkout began with precautionary stress corrosion inspection. The stage and IU were checked both visually and with dyepenetrants, as necessary, for any indication of cracks in structural members. No anomalies were found.

ASTP Launch Vehicle

The stage's eight stabilizing fins were installed and the stage erected vertically for further checkout.

Stress corrosion of fins had been discovered on the Saturn IB launch vehicle for the Skylab 4 mission in late 1973, but analysis and solution of the problem was quickly accomplished.

Using procedures developed as a result of the Skylab problem, fin-mounting brackets on the S-IB stage for ASTP were rod-peened. This process places compressive stresses in the surface of the material by pounding with a small bundle of rods, operated pneumatically like an air-hammer.

To further reduce susceptibility to stress corrosion, technicians rounded out each fin attach bolt hole, to eliminate sharp corners

The privilege of advertising articles in this publication is restricted to employees of Langley Resear Center; Articles advertised here must be offered for sale, or as otherwise advertised, without regard to race, color, religion, sex, or national origin.

The Langley Researcher, an official publication of the Langley Research Center, National Aeronautics and Space Administration, Hampton, Virginia 23665, is published every other week in the interest of its employees. Address contributions to

Langley Hosts Last Review

Miss Karen D. Brender has been selected as a nominee for the 1975-1976 NASA Career Development Fellowship Program, Education for Public Management.

Miss Brender is presently working in the Space Systems Division, Shuttle Experiments Office, as an Aerospace Technologist in theoretical simulation techniques. She has been a member of the Langley science staff since February 1963.

The fellowship in Education for Public Management is sponsored by the U.S. Civil Service Commission and nine cooperating universities: Cornell, Harvard, Indiana, MIT, Princeton, Stanford, Southern California, Virginia and Washington. It is a nine-month study program at one of these institutions.

The EPM fellowship provides an education experience aimed toward improved managerial abilities and effective accomplishment of current or anticipated agency missions. It also is meant to provide an opportunity for personal growth through intellectual stimulation, extended contacts with peers from other agencies, and time to review and integrate knowledge with personal

Cover Design: Quality Publications

Covers in this category are dramatic and appealing. The main purpose of the cover is to attract the reader and serve as an effective preview of what is to follow. It is, in one sense, a packaging problem and both front and back covers should be considered part of the package.

Concepts play an important role in quality publications and should be explored thoroughly. Special attention must be paid to the quality of the photograph or illustration as well as the content. Superior quality will help distinguish the cover from other competitive publications.

4/color process printing is desirable, wherever possible, to contribute to the general effect and prestige of the publication. While it is impossible to show our demonstrations in color, all of the covers below were designed in 4/color process.

Note that space has been used to separate the mission patch from the NASA logotype and signature.

Back cover

Front cover

Cover Design: Case Bound and Educational Publications

Case bound covers are kept simple with heavy emphasis on typography. They should project a certain classic or timeless look and cannot be over-embellished. Careful attention to production techniques will ensure a more successful overall package.

It may be advisable to use the NASA logotype inside the publication rather than on the cover—thus achieving better reproduction.

Educational publications can rely on typography in a fairly direct manner. The title is the most important consideration and can be handled in a contemporary way without pretentious frills.

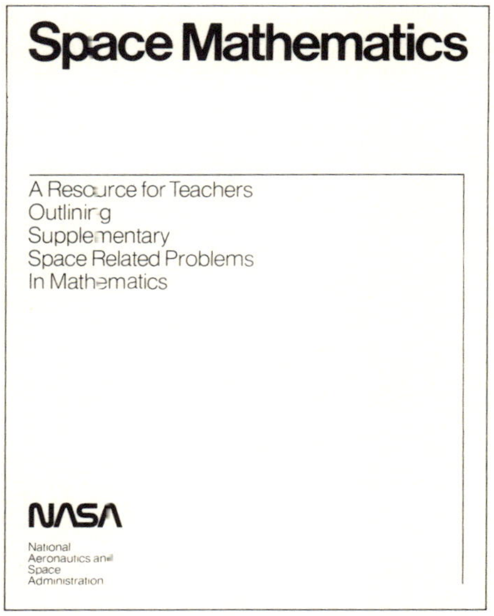

Press Kits/Directories

The main elements in the design of press kits are layout, typography and the logotype/signature. Other visual motifs are unnecessary and may be in conflict with the material inside the kit. This is an area of design which is best served if the graphics are modern but not elaborate.

The main function of a directory is to convey data and information in a concise and efficient manner. Since most directories will have very plain interiors, there is potential to apply a graphic device, illustration or photograph to the cover. One example of this is shown in the demonstrations below.

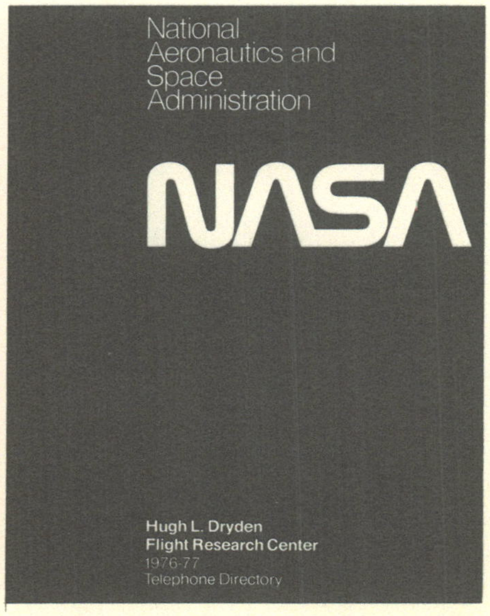

Posters/Broadsides

Posters are a vital and effective form of visual communication. Immediacy is the most important single element in poster design so it is important to avoid conflicts which will compromise the effect. Simple, bold headlines along with uncomplicated graphics can produce the desired results. Remember that the logotype/identification must be large enough and positioned well for good readership and retention.

Broadsides have similar characteristics as the poster but usually carry much more information and thus require considerable study and attention. Both mediums can have impact and are desirable tools of the communicator.

The Grid—What it is

A grid is a predetermined understructure which the designer can employ to give a publication cohesive style and character. It is a great organizer of material and as such will save countless man hours in execution. It will also help bring continuity to various diverse publications.

There are a multitude of grids which can be developed and used by the designer. In the illustration below we explore the rudiments of a grid and its application to a hypothetical publication.

1. Starting with the overall size of the publication, a designer analyzes the type of information, photographs, captions etc. which are required.

2. The grid is applied to the double-spread—it will determine all margins, gutters, folio placement, etc. In this case a 2-column grid is demonstrated.

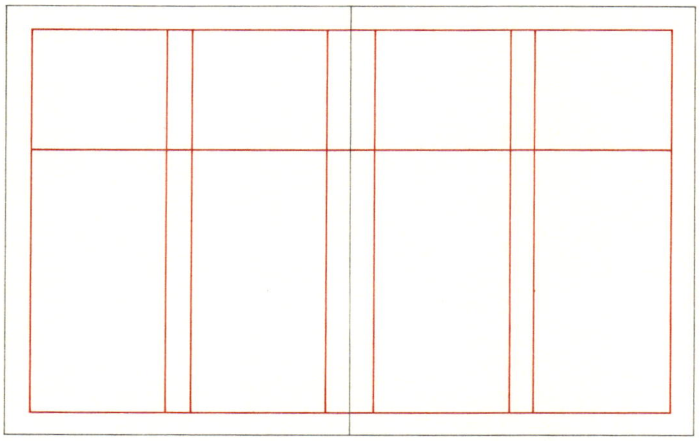

3. The designer can now begin to block in the various elements such as headlines, columns of text typography, photographs, captions and folios. This approach can be applied to the entire publication, including its cover. (Folio placement should always be in the outside margins.)

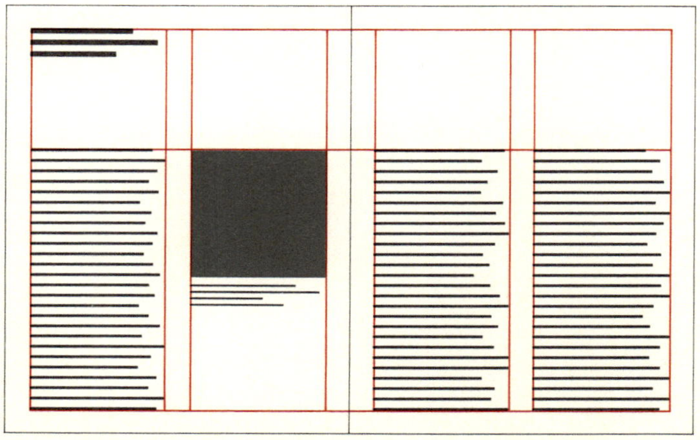

Relating Cover & Interior Formats

It is possible and desirable to use the interior grid to make a more successful cover design, one that looks like it "belongs" to the publication it houses.

It is therefore advisable to solve the publication design as a whole, rather than attempting to execute the cover out of context. Once the interior attitude has been determined it is possible to relate this back to the cover.

The examples below indicate the variety of approaches:

a. Grid based on alignment with square image on front cover. Headlines at top of page.
b. Grid which has all headlines and text "hanging" from top of page. Photographs at bottom.
c. Grid which has a symmetrical center line from which text and photographs break up and down.

a

b

c

Interior Grid Formats: Leaflets & Folders

It is advisable to use one-column formats in small brochures and folders. Grids which can be employed are numerous and three are shown below.

a. Headings align at top of page. Text and illustrations occupy major portion of page. Captions are positioned in double margin space across bottom of brochure.

b. Headings and text align at top of page. Photographs on separate track at bottom of page with captions above.

c. Headings and captions align at top of page. Text is confined to bottom half of page. Photographs are full bleed pages.

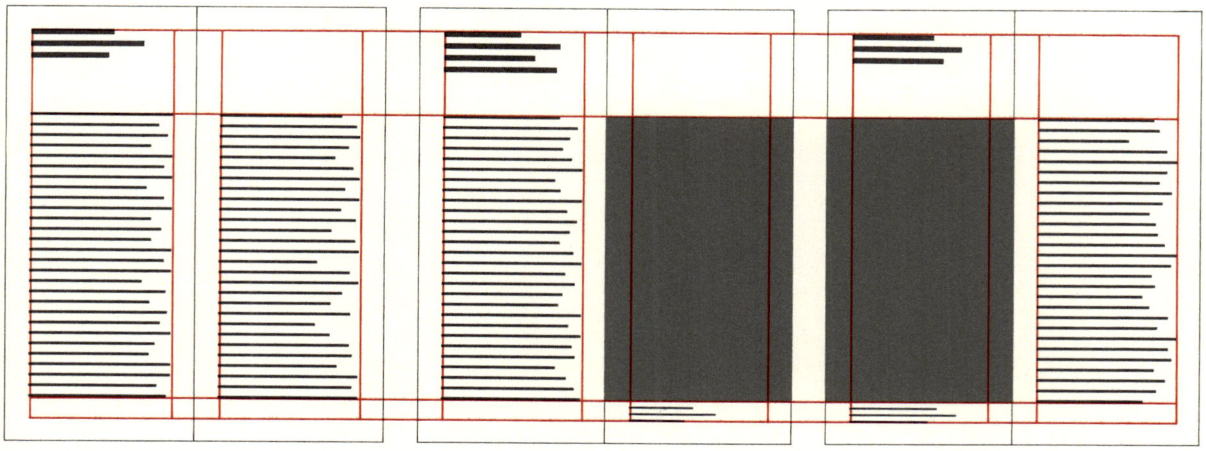

a

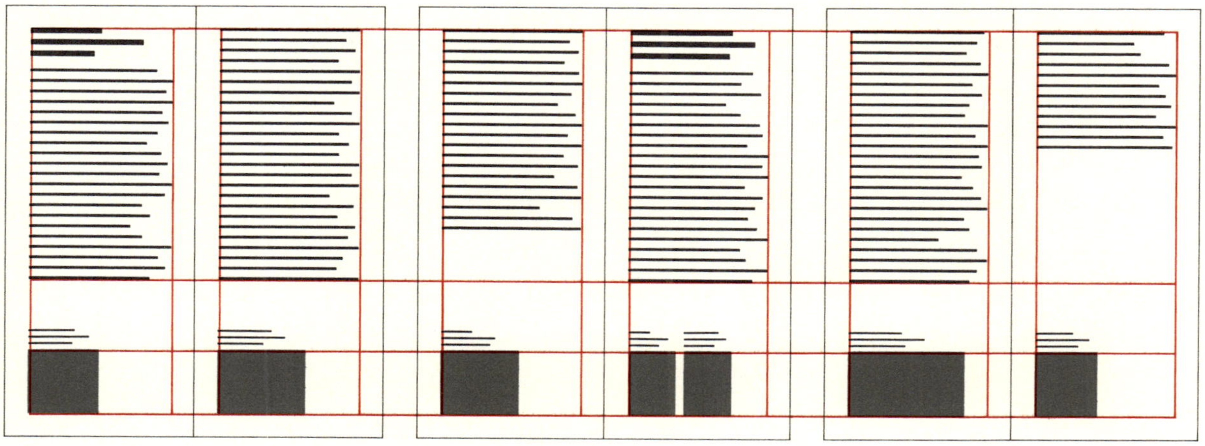

b

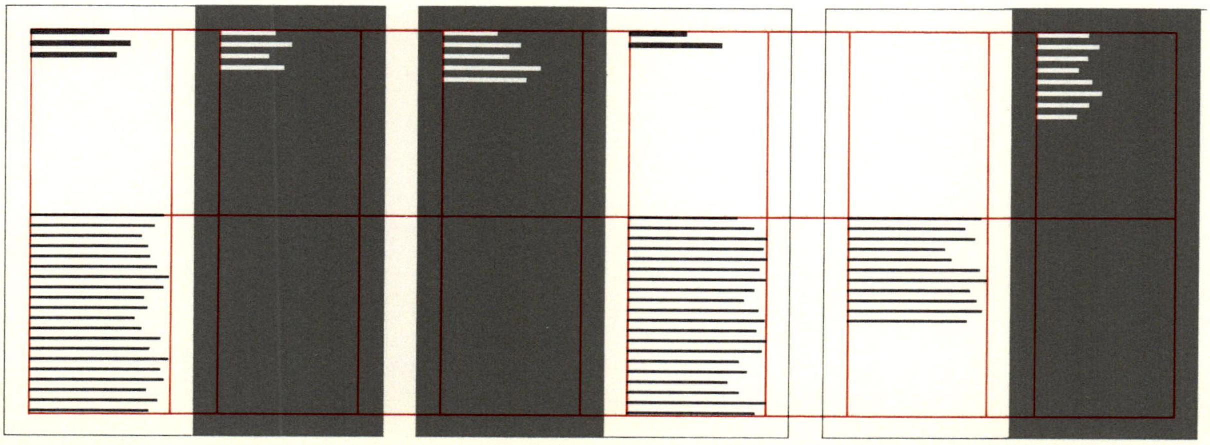

c

Interior Grid Formats: Journals and Technical Publications

In general, two-column formats will function best for this category of publications. Because of the technical nature of these publications it is best to use most of the space on the page. Some relief from the volume of type is desirable as indicated in the formats below.

a. White band across top of page functions as "breathing space." This area might incorporate folios and an occasional important headline.

b. Top to bottom use of space but with some open areas of illustration, diagrams, etc. Outline boxes should be used around all technical diagrams. Rules are employed to separate articles.

c. Top alignment of headings and text. Single page has been divided into 6 equal rectangles. Captions are positioned under photographs.

a

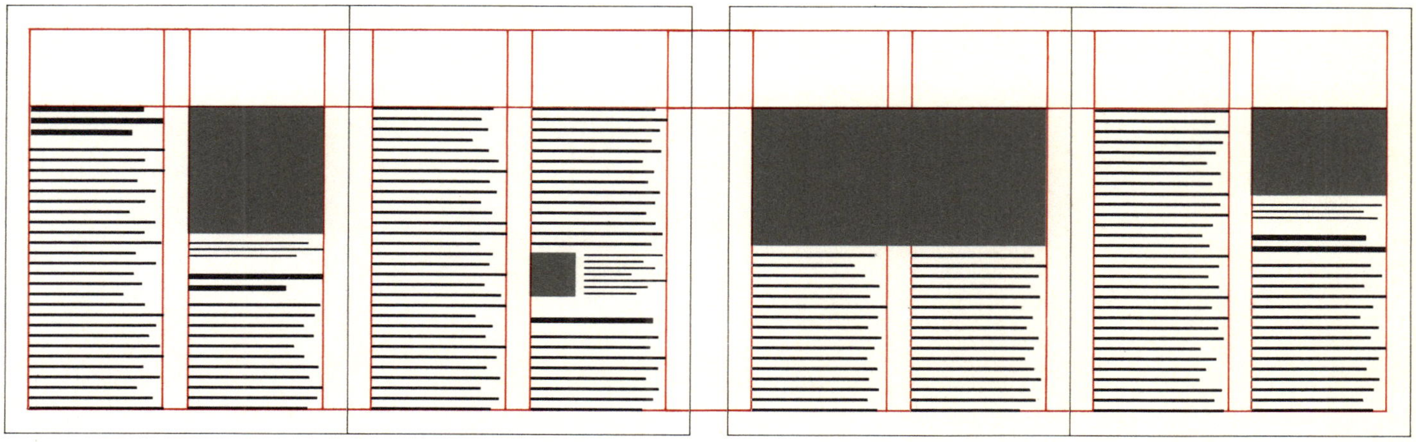

b

c

Interior Grid Formats: News Publications

The examples shown on this page are models for news publications. There are other grids which can be developed depending on the characteristics of a specific publication. Most publications of this type have a great number of articles, photographs and captions to be displayed and a good underlying grid can be very valuable.

a. Employs a two-column format. Uses a bracket device to isolate photographs and captions which are complete in themselves—not relating to specific article.
b. Uses a three-column grid. Rule motif is used to isolate photo essays. Photographs always positioned at top or bottom of page.
c. Large scale use of photography in a three-column format. Photographs positioned in loose configurations.

a

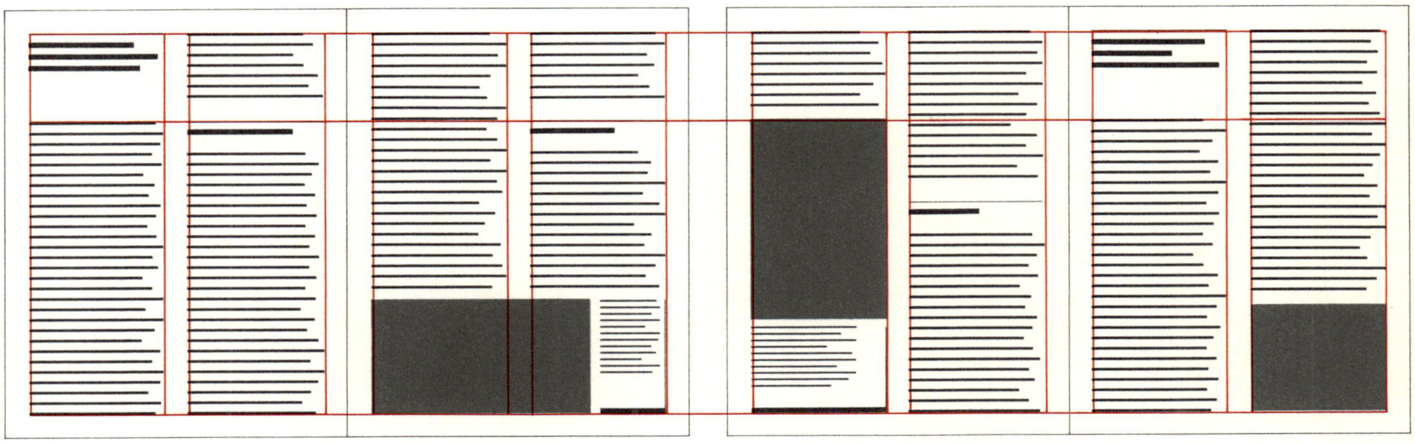

b

c

Interior Grid Formats: Quality Publications

This category of publication is perhaps the hardest to format in this manual as there are endless possibilities. Generally speaking, the formats should enjoy more open space and project a spirit of freedom rather than confinement. Three-column formats will create more movement and flexibility than two-column.

Shown below are three possible formats.

a. Large scale photographs or illustrations which bleed off left and right side of page. Typography hangs from top of page in "rag" column treatment (unequal columns). Top and bottom alignment is overall theme.
b. White border margins used throughout. Several horizontal reference lines. Text confined in block area which echoes shape of ⅔ page photograph.
c. Text occupies similar space on each page, photographs or illustrations used in small, medium and large scale. Captions for all images are grouped in one place.

a

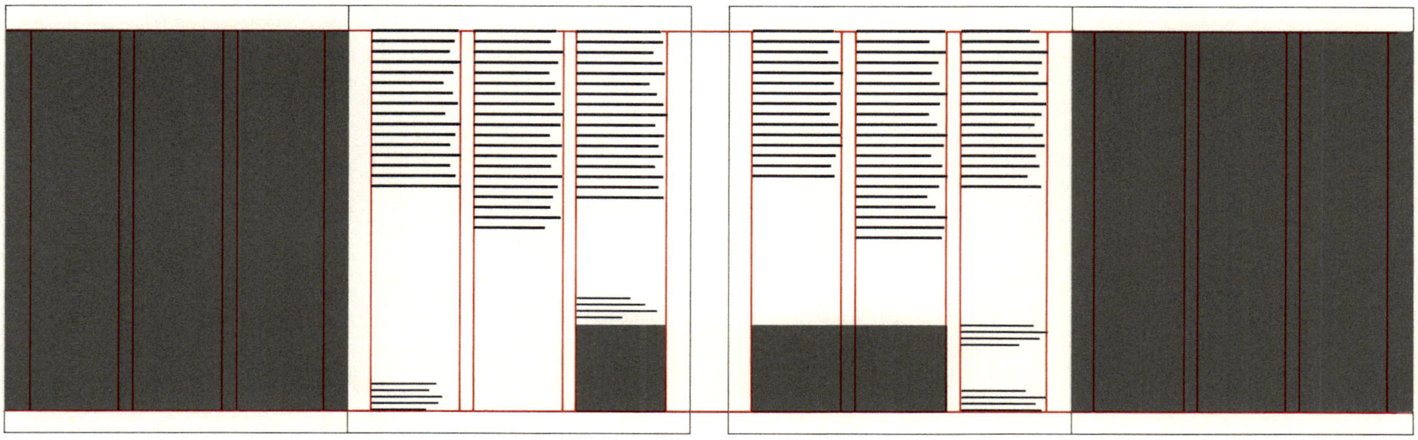

b

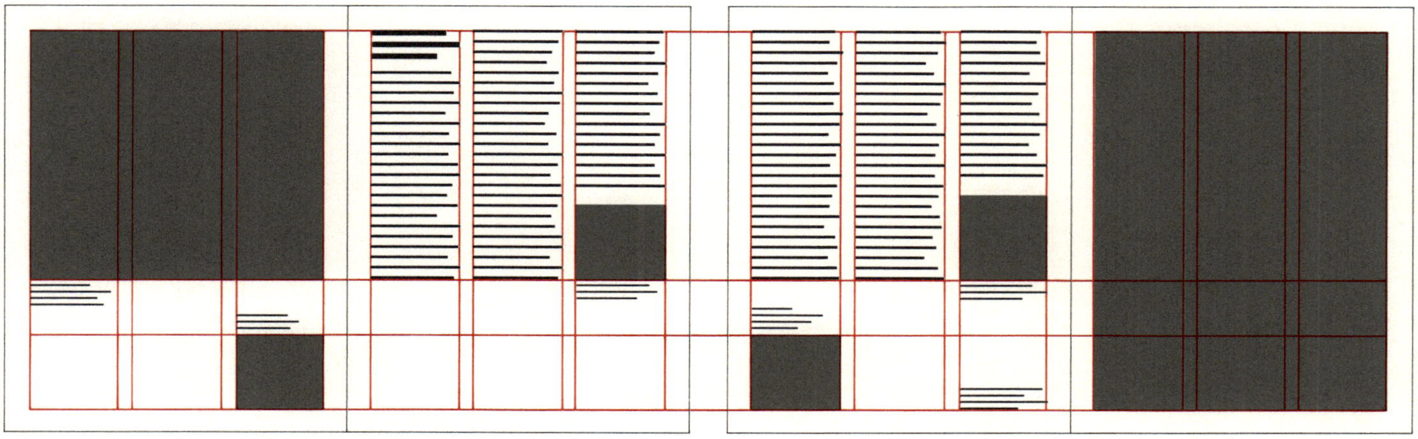

c

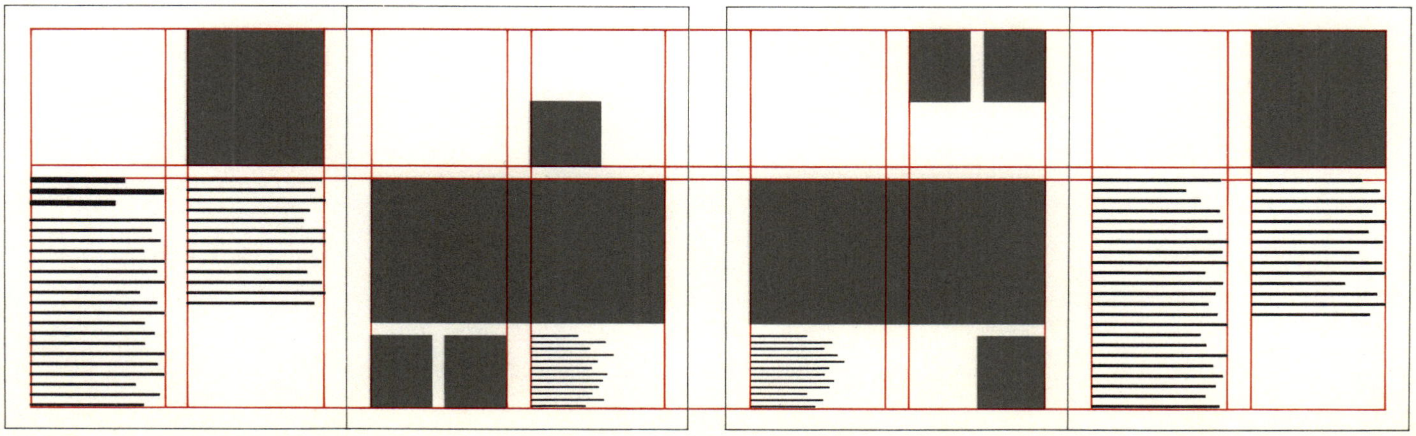

Interior Grid Formats: Case Bound and Educational Publications

These publications are of a more traditional nature and are handled in a slightly more restrained manner. Serif typefaces are appropriate here as well as the occasional use of justified typography.

a. Two-column format with wide margins. All illustrations confined to width of text typography and positioned where appropriate to editorial matter.
b. Single-column format with wide measure text setting. All visual matter is displayed in margins and is combination of square finish and silhouette.
c. Horizontal white band at top to carry headings and captions. Single-column text setting with bleed photographs. Possible use of other refinements such as large scale numerals to designate chapters.

a

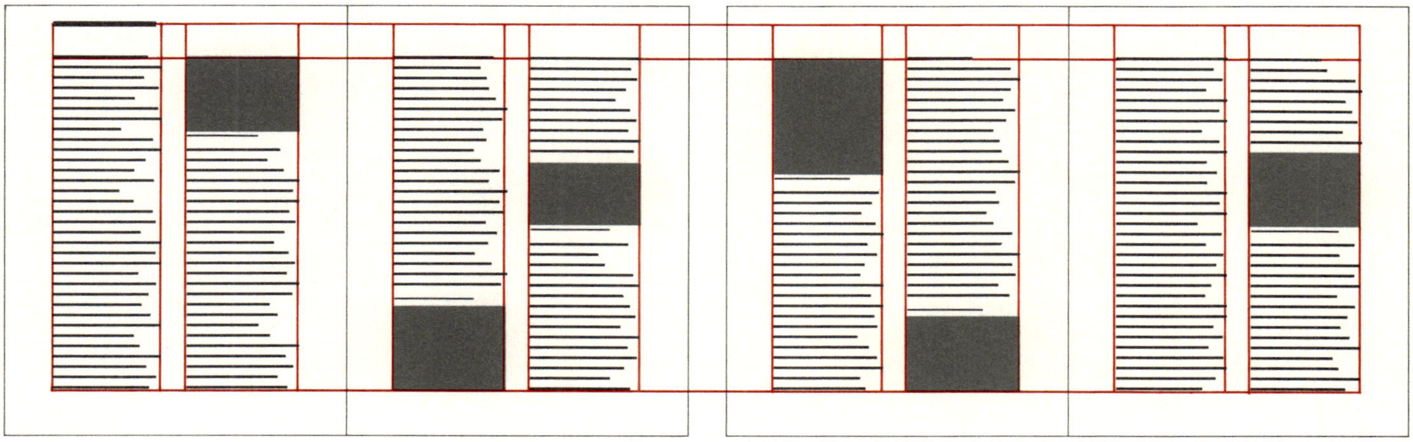

b

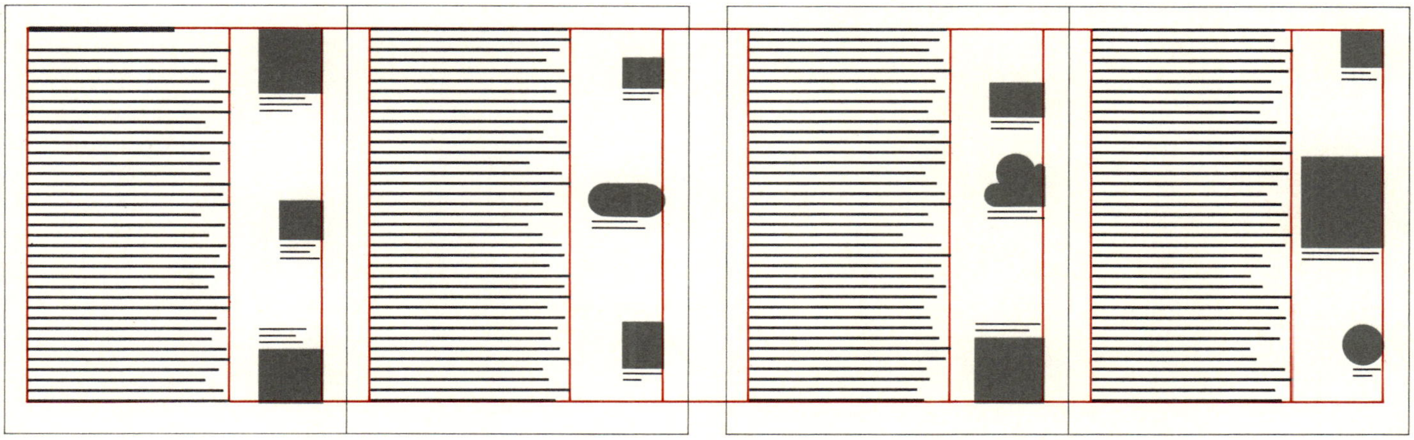

c

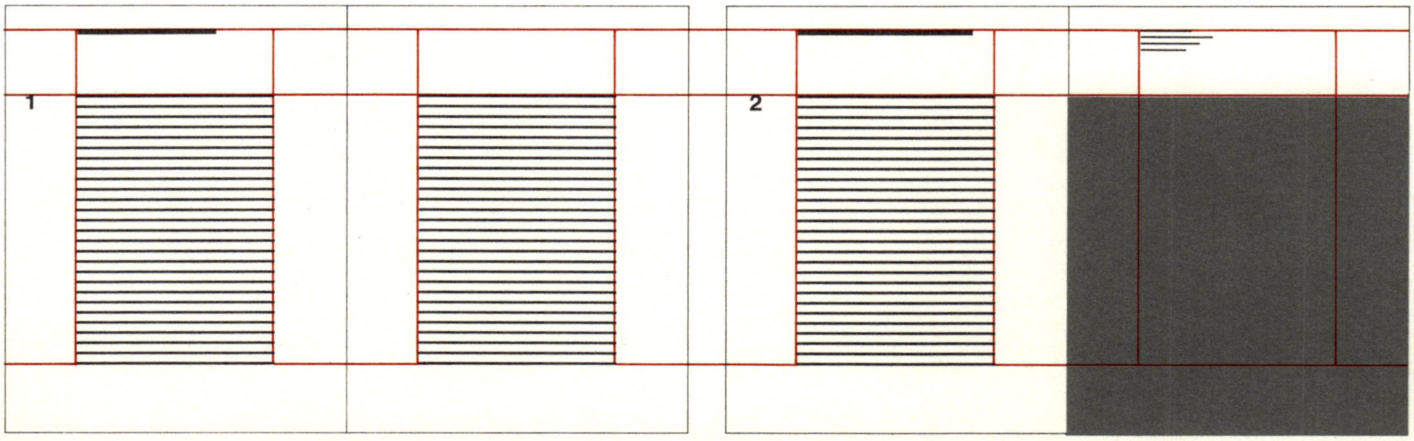

6

sequence of events which includes: arriving at a facility, going to a specific building, then seeking a floor and room number. Relating the specific sign to a larger context will yield the best results. Also, categorize signs by functional types as a method of simplifying the overall signing task.

A sign should be thought of as a large-scale headline; therefore, language should be clear and concise. Brevity is desirable in order to communicate quickly, especially to drivers of vehicles.

Placement of the sign is very important. The sign should be placed for optimum viewing distance. It is good to test these conditions by creating mockup signs out of photostats and inexpensive materials, and thus determine their effectiveness before fabricating the finished product.

Consider environmental factors when developing signage. Weather conditions should determine the material selected and the fabrication technique. Color should be chosen based on the type of Sun conditions which prevail; i.e., a dark background with reversed (white) letters will be more legible against a bright desert sky.

Use consistent message formats to create a uniform look and coordinated sign program. NASA signs should employ the flush left, ragged right format as demonstrated on these pages.

9a
Mockup and
Integration
Laboratory

b

NASA
Alpha Omega
Corporation

c

NASA
National Aeronautics and
Space Administration

John F. Kennedy Space Center

d

Tour
Information
1610 on
AM Radio

h

i

P

Tuesdays and
Thursdays

1200—1630 Hrs.

j

Gate Open

Weekdays
0700—1900 Hrs.
Evenings 2000—2130 Hrs.

Weekends and Holidays
0900—2230 Hrs.

All personnel must enter
through Gate A.

k

Warning

Hazardous Area

It is unlawful to enter this area without permission of Installation Security.

Lewis Research Center

l

STOP

Solicitation without an appointment and written permission is prohibited.

Ames Research Center

m

NASA Signs

<u>General Principles</u>

Simple, functional, contemporary signs are an integral and effective part of the NASA Unified Visual Communications System. The sign demonstrations shown on these pages should serve as models when signs are being developed for a particular site, building, or facility. They are intended to provide general guidelines when NASA sign systems are being planned.

Signs function on many different levels, but their basic purpose is to communicate to a specific audience. They identify facilities, guide to a desired location, warn, notify, or announce something to the sign user.

This signage section is divided into two fundamental parts, exterior (6.1) and interior (6.2), but certain principals apply to both categories. The following points should be reviewed and considered at the inception of signing activity.

Employ a systems approach to signing. Begin by developing an overall plan of signage based on a logical

Exterior Identification

a. Site sign at major facility. Should be executed in a permanent material such as aluminum or molded reinforced fiberglass. Should include a reveal (shadow groove) between sign and posts.
b. Building mounted identification sign.
c. Contractor sign. Possibly modular so that the bottom portion can be replaced.
d. Wall mounted, free standing metal letters. May be appropriate for a main building or Visitor Center. Dimensional letters must be of highest quality and complementary to the architectural surface. Mockups should be tested for material, color, height and depth of letters, method of affixing to building.

a

Exterior Informational

e. Wall or ceiling mounted directional signs. May be metal or acrylic. Note: relationship of arrow to height of capital letter.
f. Facility directional sign constructed similarly to item a (above).
g. Trailblazer signs. Can be metal for permanent use, or painted wood for temporary use.
h. Informational sign uses abbreviated language style for speed of communication.
i. Map and location directory can be metal, reinforced fiberglass, or other permanent medium. Graphics must be simple and stylized to aid the user. May be back lighted if appropriate.
j. Parking sign with replaceable bottom portion. Letterform "P" is from D.O.T. Symbol/ Signs library.
k. Sign displays much information in a small space.
l. Modular sign serves to warn viewer of possible danger and restricts access.
m. Modular sign restricts entry or passage.

e

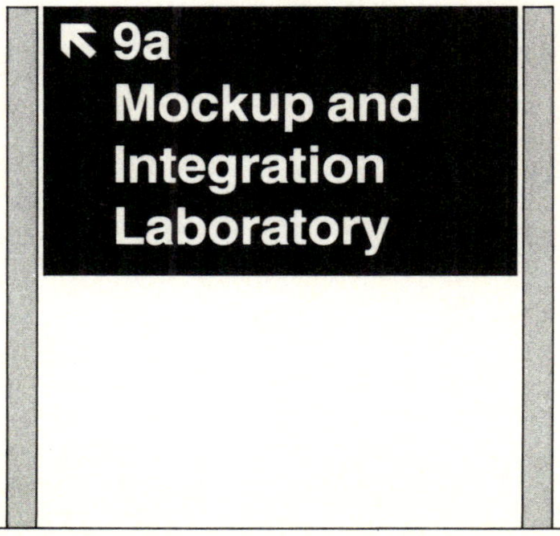

f

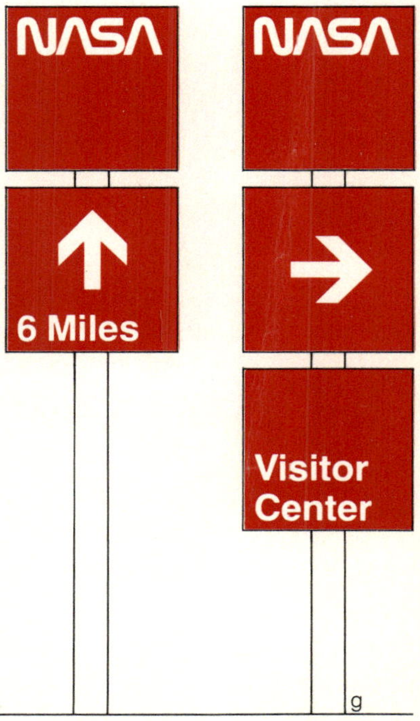

g

which serves its basic utilitarian function while complementing the architectural scheme.

Consider flexibility when designing sign systems. Often a sign which has a modular capacity can be very economical in the long run. Several of the sign models shown here are modular in order to have a greater life span and be less expensive.

Choose materials according to longevity requirements, permanent versus temporary, and consider the cost-effectiveness of the material. Sometimes the material which seems to be a bargain does not turn out to be if the sign must constantly be replaced or refurbished.

Incorporate the NASA logotype where practical and appropriate. Site signs, temporary and permanent, should include NASA identification. Major buildings and Visitor Centers should be marked as well. This will create maximum exposure, and will strengthen the overall perception of the Agency. Use the grid drawing of the logotype (1.7) for large-scale use on signs.

Consider mounting techniques, especially for interior signage, which are simple and flexible; i.e., small plaques and signs can be affixed with foam tape which does not mar the surface.

In general, try to coordinate signage needs so that a number of signs can be produced at the same time. This will produce a better, more effective family of signs and will also be more cost-effective to the Agency.

e

f

g

i

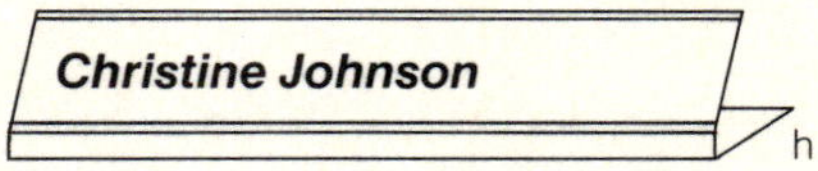

h

l

m

n

NASA Signs

<u>General Principles—Continued</u>

Helvetica, the keystone typeface in the NASA Unified Visual Communications System, is used on all exterior and interior signs. The models shown in this signage section feature Helvetica Medium. Occasionally, such as on building directories, Helvetica Regular is used in combination with the Medium weight. Set large words in Display phototypography, optically spaced, and camera-enlarged to needed size. Remember to use "normal" rather than "tight" letterspacing for signage typography. Avoid hand lettering or other crude execution techniques.

Keep the signs simple. Avoid the use of borders or other types of artificial embellishment. These only serve to clutter the sign and lessen its effectiveness as a communication tool.

Signs should always be designed to integrate with the architectural setting. Choice of material, color, and size should be based on an awareness of the environment where the signs will be viewed. Effective signage is that

Interior Directional

a. Ceiling mounted sign which is modular, allowing sections to be replaced or transposed. Should be of permanent material and integrate with the existing ceiling hardware.
b. Wall mounted version of item a.
c. Similar to items a and b, but with a different color scheme and language.
d. Floor directory is modular and made of a permanent material like acrylic or metal. May be rear illuminated if appropriate.

Interior Identification

e. Wall plaque of acrylic with subsurface screen printing (on back side) for permanence.
f. Door plaque of acrylic with sliding panel for easy replacement.
g. Wall mounted or door mounted plaque with sliding panel for occupancy functions.
h. Desk identifier with replaceable name.
i. Wall mounted plaque with replaceable name.

a

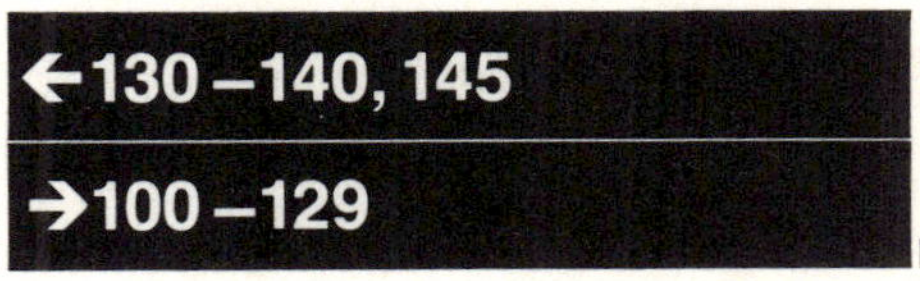

b

c

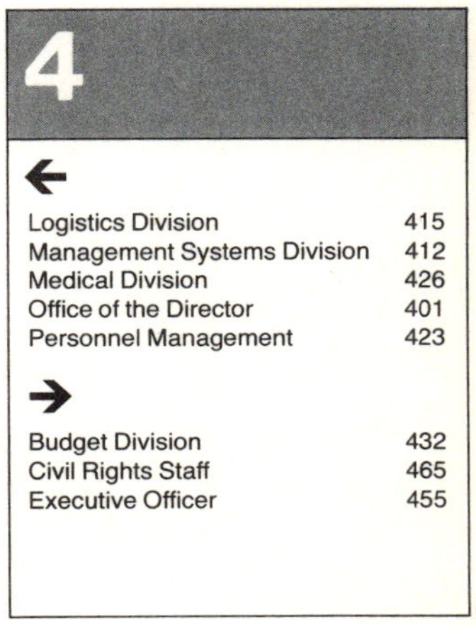

d

Interior Informational

j. Building directory is modular and made of metal or acrylic. Could be wall mounted or free standing. Rear illuminated if appropriate.
k. Movable sign for temporary use. Constructed of metal with replaceable message panel.
l. DOT Symbol/Signs employed rather than unnecessary words.

Interior Regulatory

m. Symbol/Sign restricts smoking activity.
n. Wall or fence mounted sign warns and restricts access to certain areas.

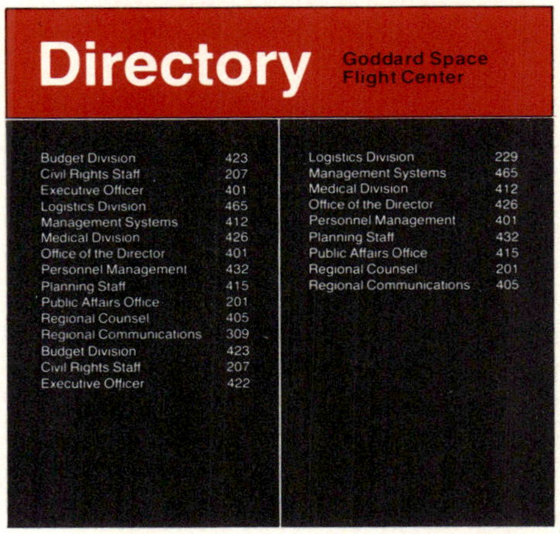

j

k

7

NASA Motor Vehicle Identification

The identification markings of all NASA owned or leased vehicles conform to GSA regulations regarding type sizes and the positioning of the identification on both front doors of the vehicle, or in a prominent location on the vehicle when the former is impractical.

Four elements comprise the standard vehicle identification configuration: the government use identification (¾″/1.9 cm cap height), the NASA logotype (3⅛″/7.9 cm cap height), the agency identificaton (1″/2.54 cm cap height), and the installation identification (1″/2.54 cm cap height).

The identification configurations are pre-spaced, dry-transfer, mylar lettering sheets. Because of the limited door-width of many vehicles, some installation names have been set up in two lines to fit within these restrictive spaces.

As a general rule, the identification configuration is applied on a flat surface as high as possible on the front doors of the vehicle avoiding sculptured edges, door handles, side-view mirrors, and other hardware. The configuration is visually centered laterally from the edges of the door, in the same approximate position on both doors.

Due to the wide variety of vehicles used by NASA, it may be necessary to make some adjustments in the application of the agency identification to accommodate chrome strips, door guards, etc. It is imperative in these instances to maintain the flush-left handling of the logotype and all typography.

In appropriate instances where space allows, as shown in various demonstrations on the following pages, the logotype may be separated from the identification configuration and used in a larger size elsewhere on the vehicle for a better display.

Most vehicles are marked in one of three color schemes: 1) black logo and typography on all light color vehicles up to a 40% value on a gray scale, 2) white logo and typography on all dark color vehicles darker than a 40% value on a gray scale, or 3) NASA red logo and black typography only on *white* vehicles.

Shown to the right is a 10-step gray scale showing values ranging from a 10% gray to a solid black. When marking vehicles where the color of the vehicle is in a middle-value range and it is uncertain whether the identification should be white or black, place this scale up to the color of vehicle in question to determine the value. The edge of one step on the gray scale should nearly disappear in juxtaposition with the color of the vehicle.

"Special" vehicles that require painting and offer high visibility to the general public (i.e. tractor-trailer vans, buses, etc.), should be painted in the white/blue scheme with a NASA red logotype, as shown on the following pages. Proposed designs for such vehicles should be submitted to the NASA Graphics Coordinator.

Two sizes of typography are available for application to large vehicles. For tractor-trailer vans (and vehicles of a similar scale), the cap height of the typography is 3½″. For buses (and vehicles of a similar scale), the cap height of the typography is 2¾″. The agency name always remains in the two-line configuration and the center identification appears in one line positioned below the agency name at a distance of two-times the cap height, baseline to baseline. For example, on a tractor-trailer van, the cap height of the typography is 3½″ and the center name is positioned 7″ below the second line of the agency name, baseline to baseline. When 2¾″ cap height type is used, the center name is positioned 5½″ below the second line of the agency name, baseline to baseline.

An intermediate size of 1¾″ cap height typography is also available for intermediate size vehicles (tractors, trucks, vans, small buses, and vehicles of a similar scale) to be used with a 5″ or larger logo.

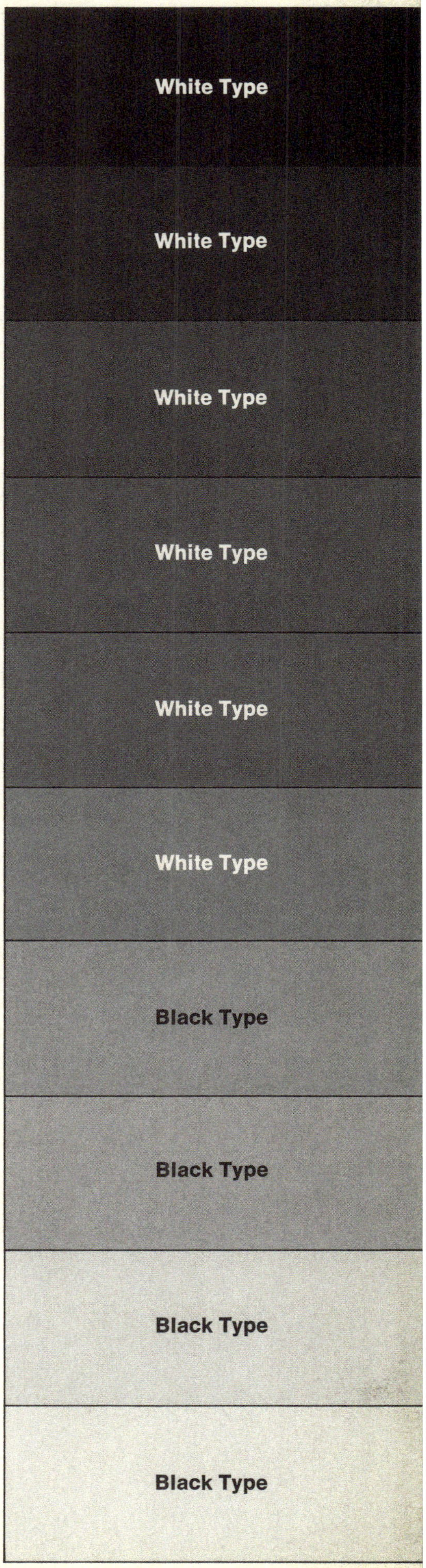

Motor Vehicles:
Various Colors & Values

The majority of NASA vehicle identification consists of marking a wide variety of makes, models and types of vehicles painted in an equally wide variety of colors and values. These vehicles are marked with either black or white identification configurations, as shown on the two sedans below.

Using the gray scale on the previous page, the color value of the vehicle is determined. If the vehicle is a 40% value or lighter, the identification configuration is black (fig. a). If the vehicle is darker than a 40% value on the gray scale, the identification configuration is white (fig. b).

Note that in these demonstrations the typography does not overlap the sculptured edges or molding on the doors and the flush-left handling of all the elements is retained.

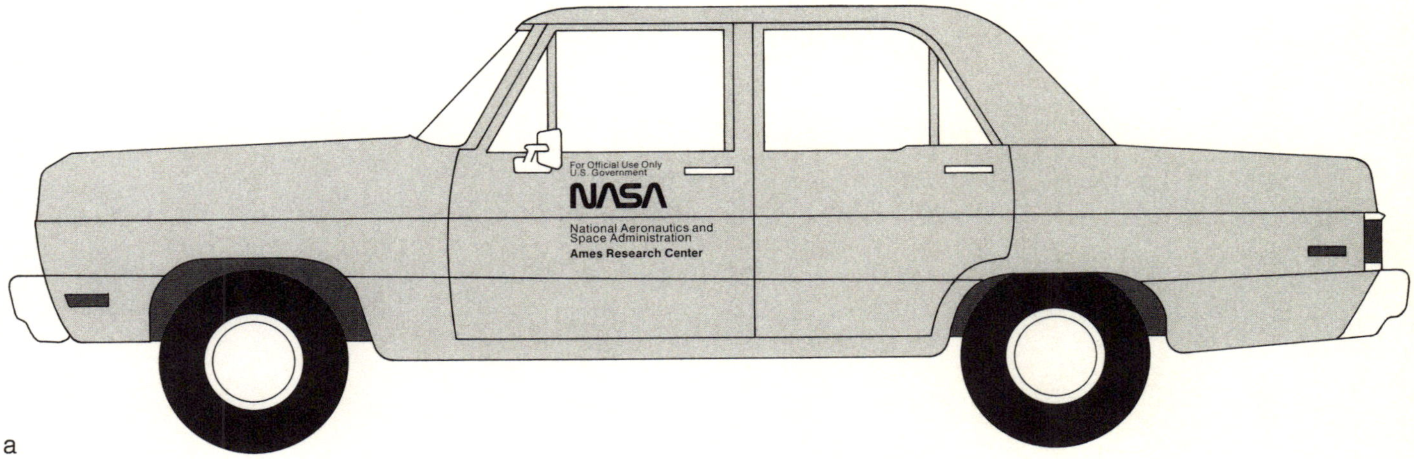

a

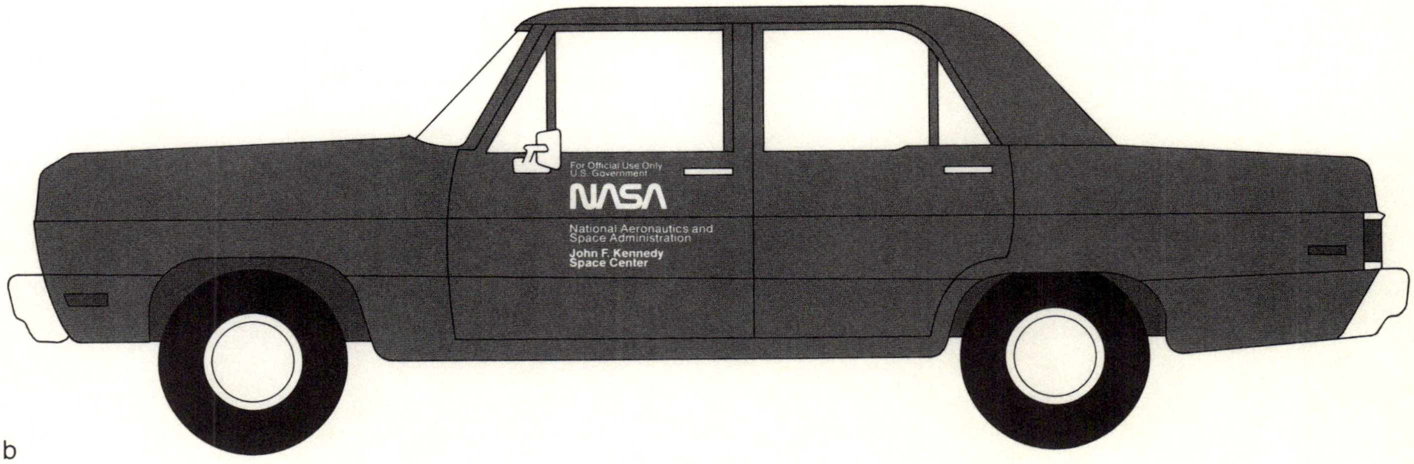

b

Motor Vehicles: White & "Special" Vehicles

A white vehicle provides an opportunity to utilize the logotype in NASA red with all other typography in black, as shown below (fig. a). The sizes and relationships of the identification configuration remain the same. Note that a red logo is used only on a *white* background for vehicles.

A "special" vehicle, intended to be more conspicuous to the general public, may be painted in the white/blue scheme shown below (fig. b). The vehicle is visually divided in half into a white stripe on top and a blue stripe on the bottom. (The specific contours of the car model often suggest where the division should be). The government use identification is black, the logotype is NASA red, and the agency and center identification is white. The flush-left handling of all elements is retained.

The color specifications for the white/blue scheme (gloss enamels from the Federal Standard Color System) are as follows: White # 17886, Blue # 15102.

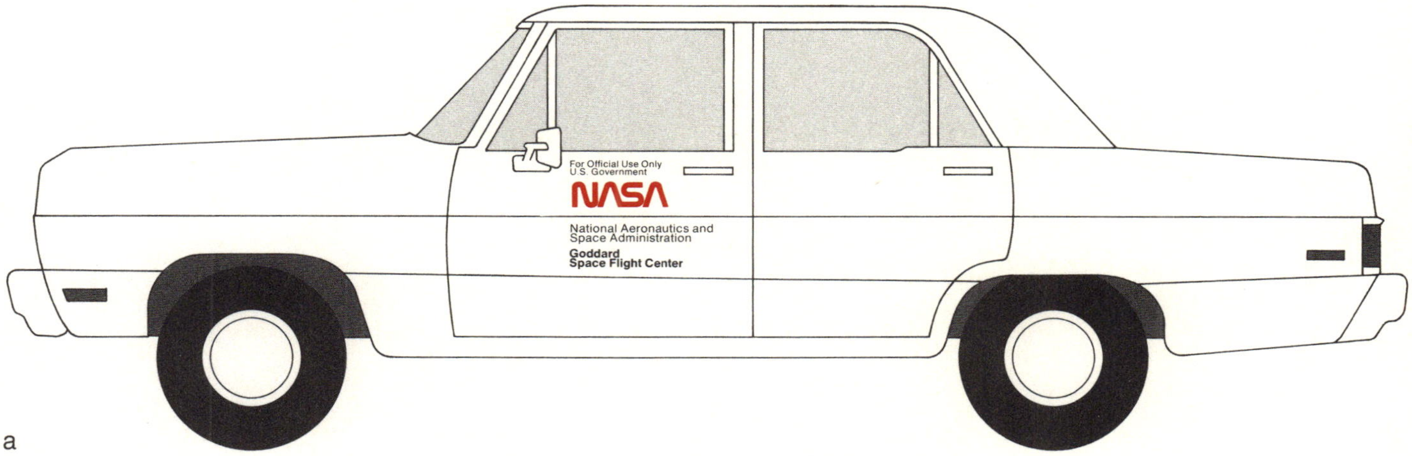

a

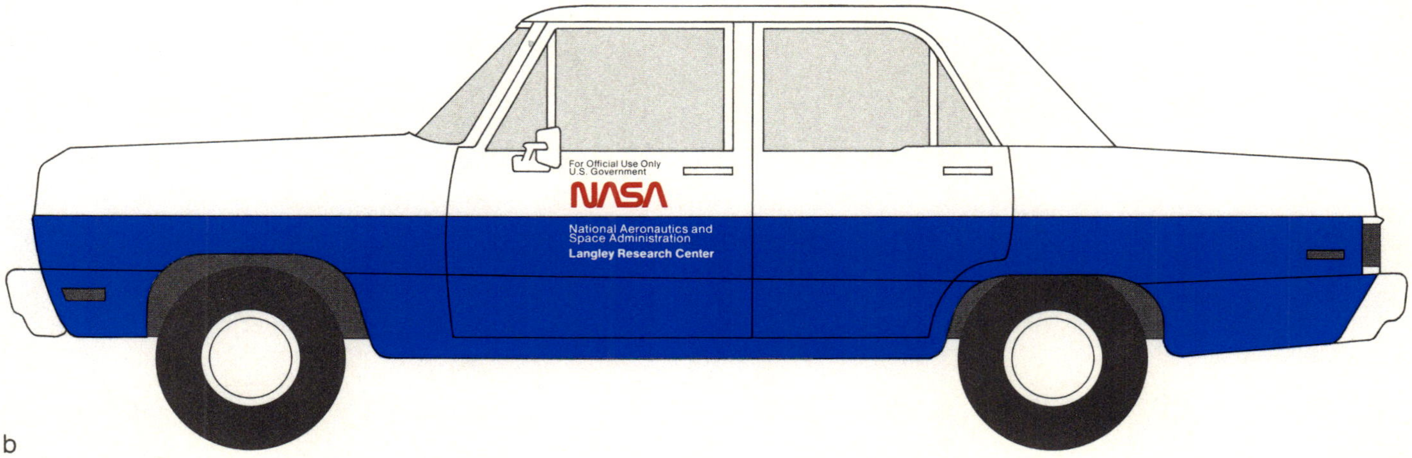

b

The Identification Configuration With the Logotype

The illustrations below show the various sizes of the four elements that comprise the identification configuration and the approximate placement of the configuration on the door of the vehicle.

These marking configurations are prespaced, dry-transfer mylar letters. The center name, prepared in either one or two lines, is positioned 2″ below the agency name as shown below. It may be necessary to adjust the positioning of the configuration to accommodate chrome molding, door guards, etc. that vary on most vehicles. It is imperative to retain the flush-left handling of all elements.

The small illustration below demonstrates the correct application of the configuration to the passenger door.

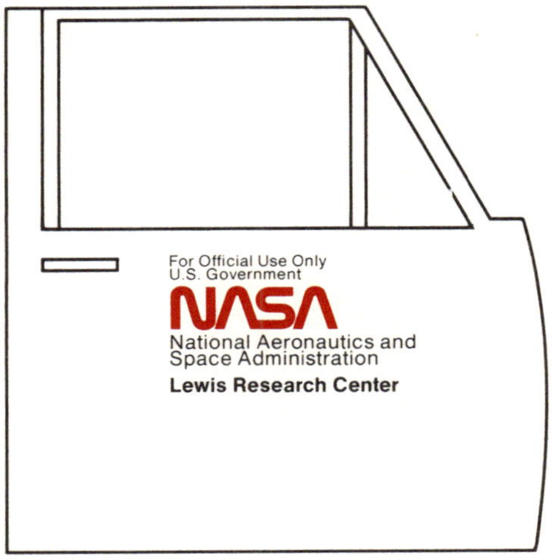

Passenger door application

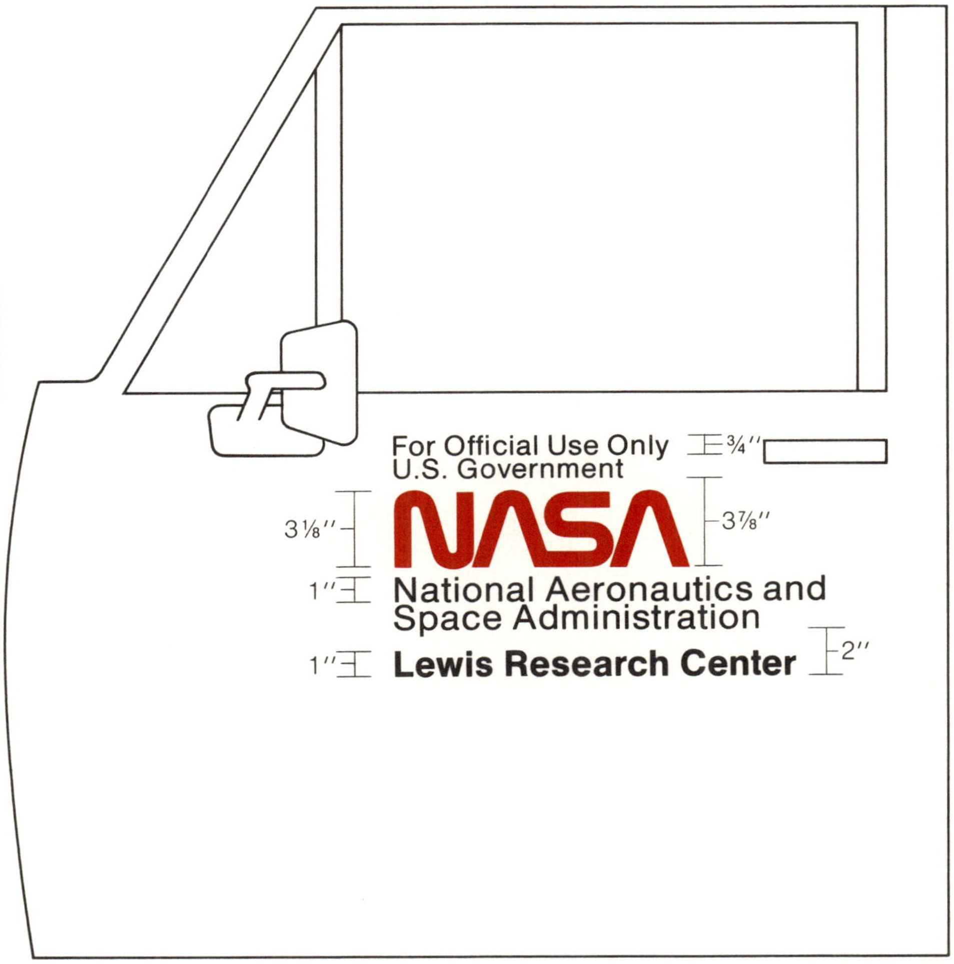

The Identification Configuration Without the Logotype

In appropriate instances, the NASA logotype may be separated from the identification configuration and positioned elsewhere on the vehicle or in a larger size. The remaining identification elements remain on the front doors of the vehicle.

The illustrations below show the various sizes of elements that comprise the identification configuration, sans the logotype, and the approximate placement of the configuration on the door of the vehicle. The 3⅛" cap height logotype appears elsewhere on the vehicle for a better display. Refer to the various demonstrations on the following page.

The small illustration below demonstrates the correct application of the configuration, sans the logotype, to the passenger door. All elements are flush-left.

Passenger door application

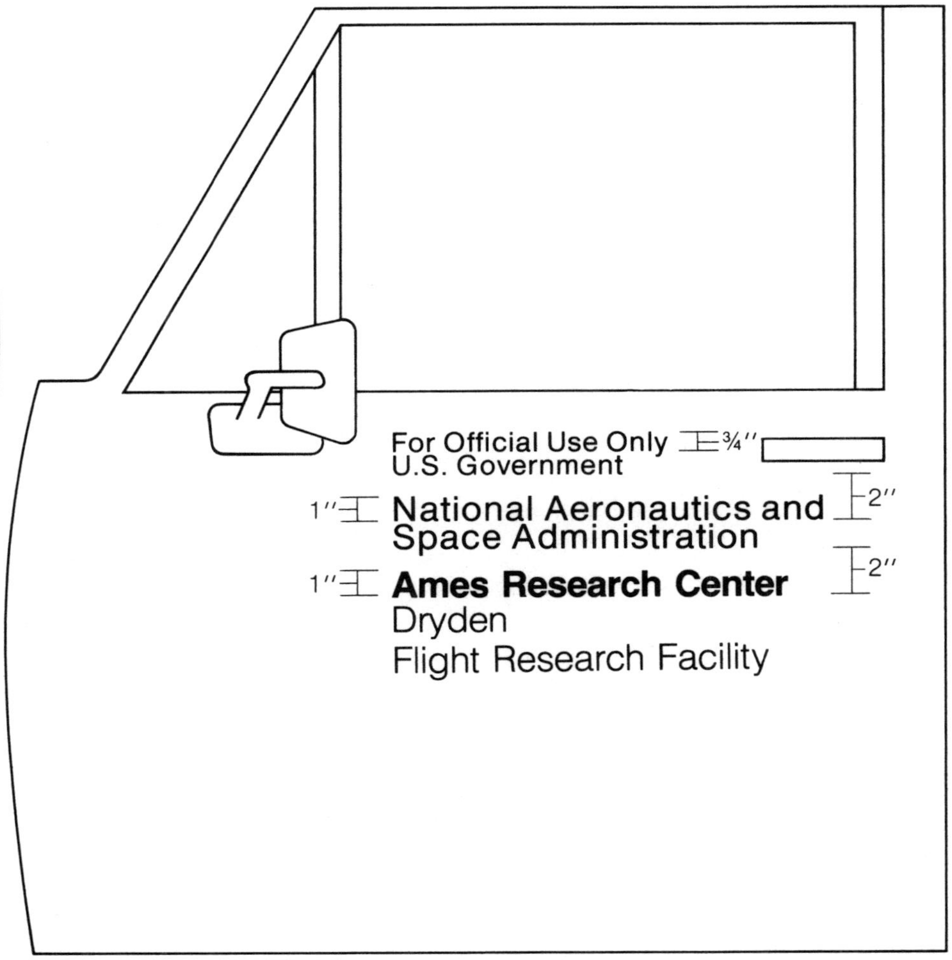

c) Panel truck with windows: The larger logotype (8″ cap height) is separated from the other identification and aligns with the x-height of the first line of the configuration. All identification is white on a dark color vehicle.

d) Tractor-trailer van: This is an appropriate use of the white/blue scheme on a vehicle which may be frequently seen by the public. The cap height of the typography is 3½″ and the cap height of the logotype is 24″ on this and other vehicles of this scale. An 18″ logotype is available for smaller vans.

e) Bus: The white/blue scheme may also be appropriate here. The cap height of the typography is 2¾″ and the cap height of the logotype is 12″ on this and other vehicles of this scale.

f) Truck: On trucks that require painting, the cab is white and the remainder of the vehicle is NASA warm gray. A standard configuration for a white vehicle is applied to the cab doors. The gloss enamels are from the Federal Standard Color System: White # 17886, Gray # 16165.

g) Station wagon: The standard size NASA red logotype (3⅛″ cap height) is positioned at the rear of the white station wagon. The typography on the door retains the flush-left handling of the elements.

c

e

g

Motor Vehicle Identification Demonstrations

The flexibility and variety of the NASA vehicle identification system is demonstrated in the seven illustrations below. The consistency within the system is achieved through the simple, clean application of the typography and the distinctive display of the logotype on the various vehicles.

a) Pick-up truck: On a color of light value, all identification is black. A 5″ cap height logotype is used at the rear of the vehicle for a more prominent display. The typography on the door is adjusted to accommodate the contour edge of the vehicle while retaining the flush-left configuration.

b) Panel truck: On a color of dark value, all identification is white. Adequate space allows the logotype to be separated from the other identification and displayed in a larger size (8″ cap height).

a

b

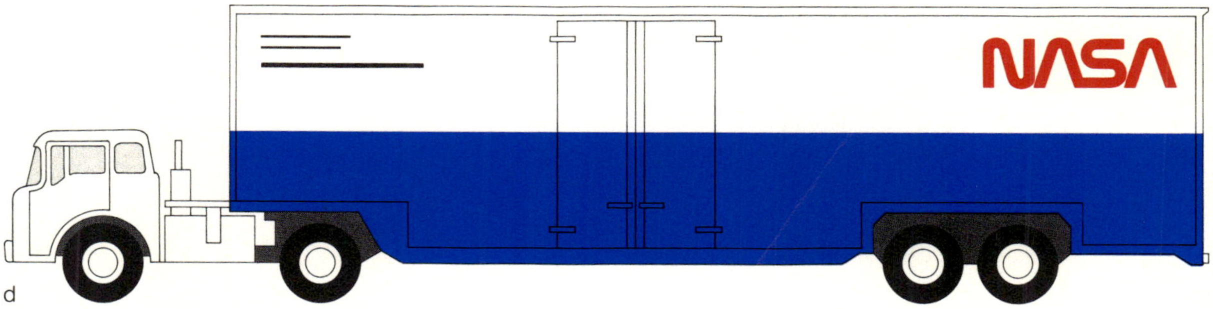

d

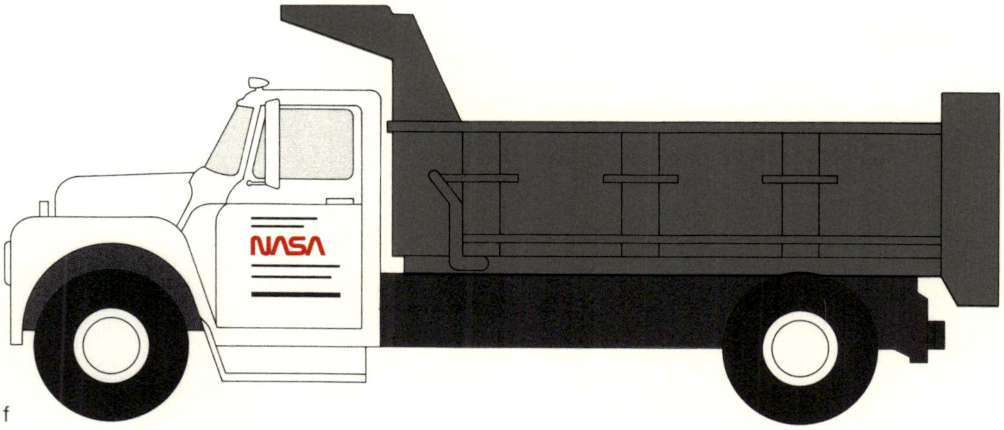

f

Aerospace Education Unit

This is an example of a special purpose vehicle that offers wide exposure of the NASA Unified Visual Communications System to the general public. The design utilizes NASA red and warm gray with a large logotype.

While this specific vehicle design is distinct from the other vehicles shown on the previous pages, it has a similar "family" look due to the use of the Helvetica typefaces, the flush-left handling of the typography, the large display of the logotype and the relationships between the various elements.

A clean, simple look is desired in all such instances, emphasizing the logotype as the primary element identifying the vehicle to the agency. The name of the specific vehicle (i.e. Aerospace Education Unit) is on the second level of emphasis. The full agency and center name are in a small scale and subordinate to these two elements. The government use information remains on the front doors of the vehicle.

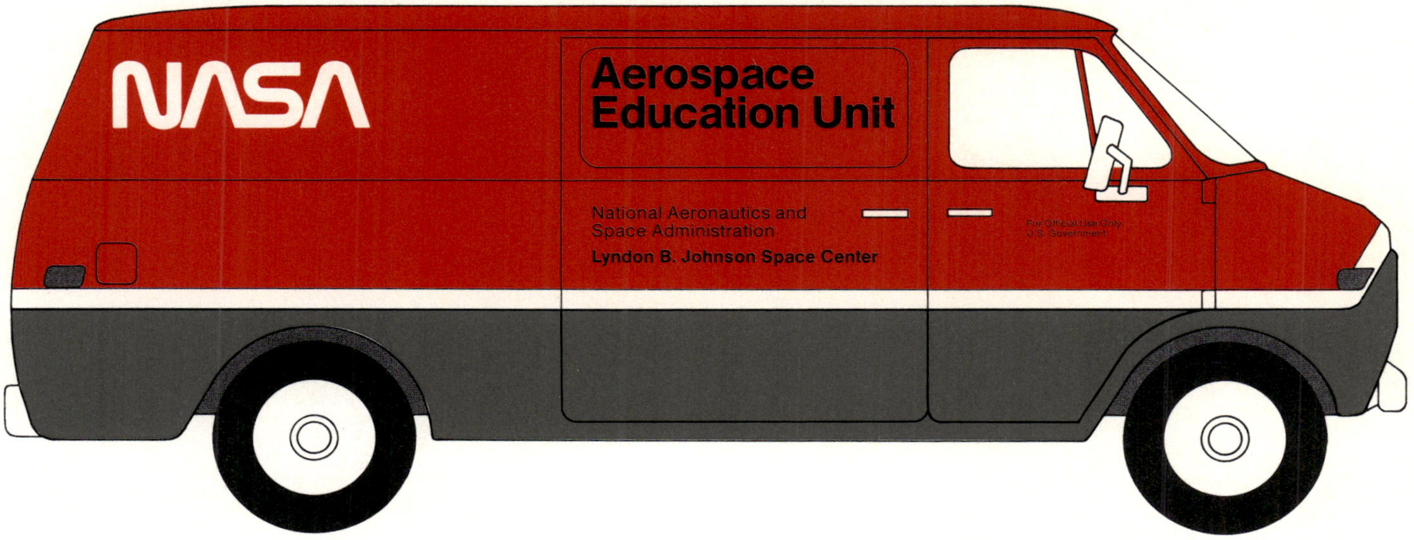

of a rubber boot or the leading edge of the vertical fin.

On the left side of the fin, the logotype and numerals are always flush left. On the right side, the configuration is always flush right in approximately the same position. When movable control surfaces are involved, the sizes and positions of these elements are carefully determined to avoid breaking the logotype on a curved stroke on either side.

To comply with FAA regulations, the complete registration number will usually appear in 12″ white Helvetica Medium within the blue stripe on the fuselage between the end of the wings and the leading edge of the horizontal stabilizer.

Exit and rescue markings shall be in accordance with U.S. Air Force T.O. 1-1-4, except that FAA regulations for commercial air carriers may be used for NASA administrative aircraft.

Aircraft Operations personnel at each field installation can provide other operational or safety marking requirements to be included in developing applications of this paint and marking scheme to other aircraft.

All new applications of the paint and marking scheme must be submitted to the Aircraft Office for Headquarters approval.

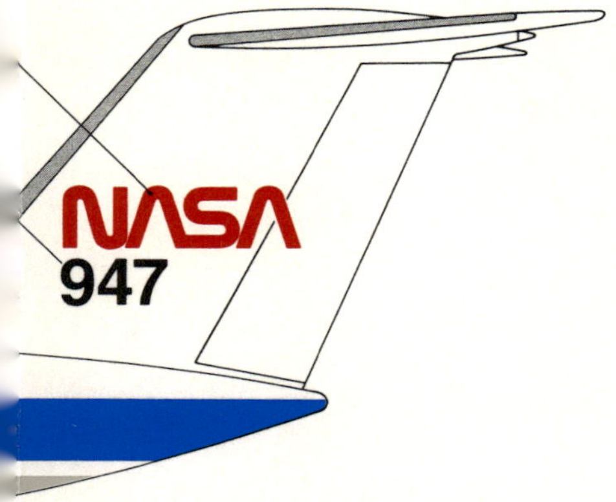

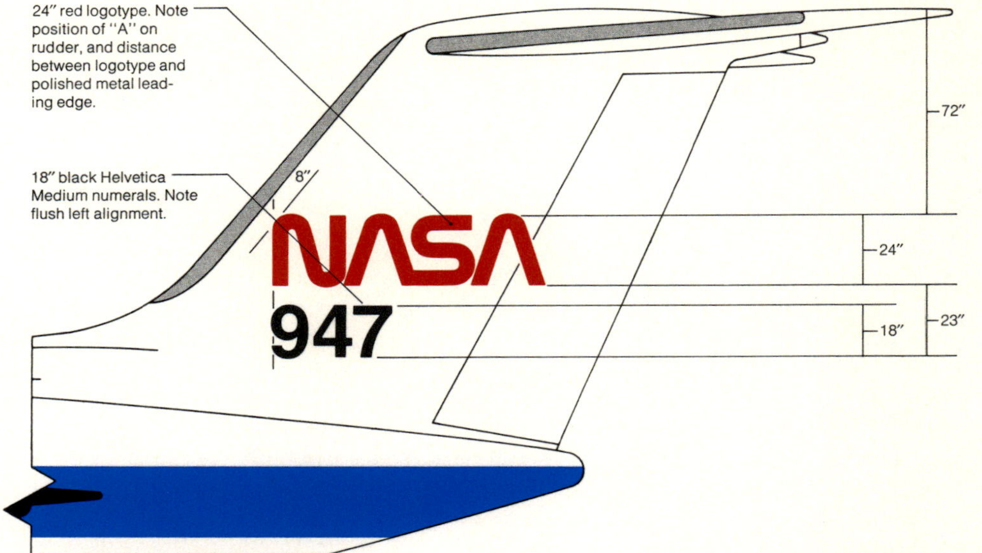

Detail A:
Left Side Tail
Scale: 3/16″=1′

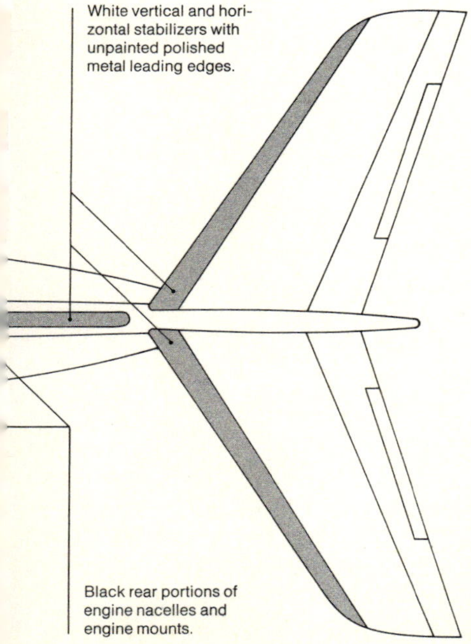

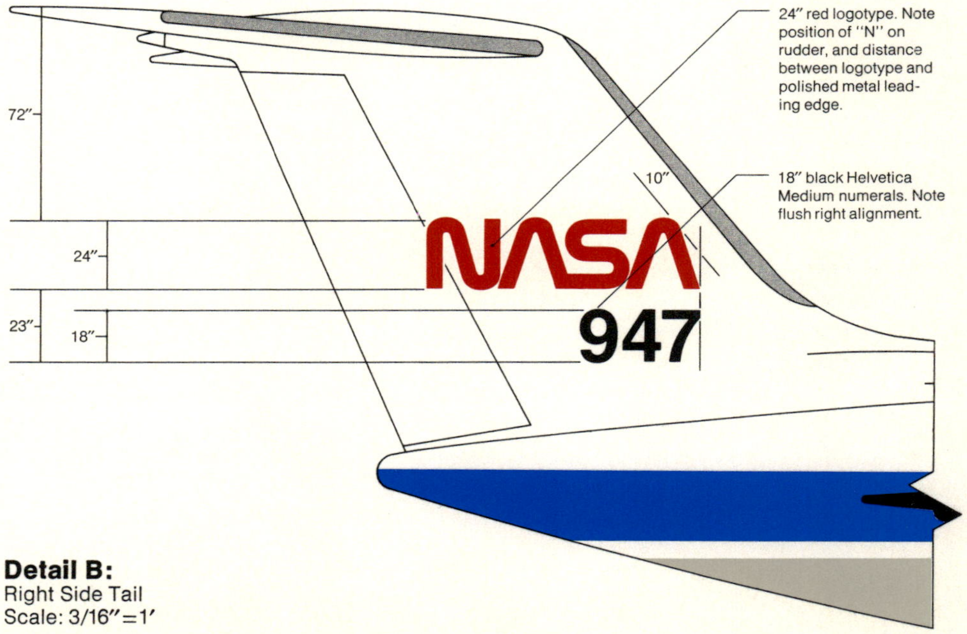

Detail B:
Right Side Tail
Scale: 3/16″=1′

NASA Aircraft Paint and Marking Scheme

The overall aircraft paint and marking scheme normally consists of a blue stripe down the middle of the fuselage sides, white fuselage top, white wings, gray fuselage bottom, FAA registration number and other markings for exits, rescue, warning areas, etc. This scheme incorporates uniform logotype markings applicable to NASA-controlled aircraft.

Some typical approved paint scheme applications are shown on the following pages. Although not complete in all details, these examples demonstrate some common relationships between the aircraft structure and the marking scheme, incorporating the following guidelines to be used in developing paint scheme applications for other aircraft:

The preferred location of the logotype is on the vertical fin (for vertical fin aircraft), unless operational considerations preclude such location. The logotype is NASA red, accompanied by the aircraft number in black Helvetica Medium. These markings are as large as possible. When the markings exceed 12″ in height, the logotype is a minimum of 25% larger than the numerals.

Based on the size logotype being used, there is approximately one stroke-width's space between the logotype and the numerals, and also between the logotype and the edge

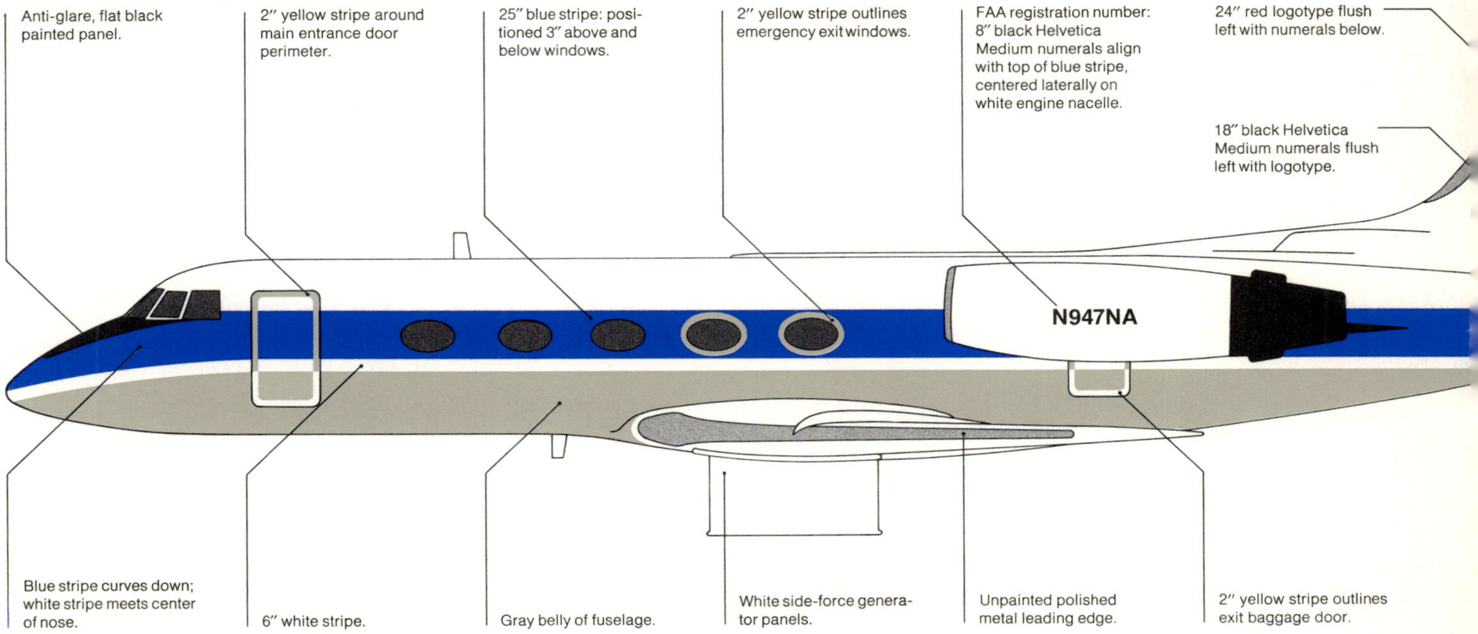

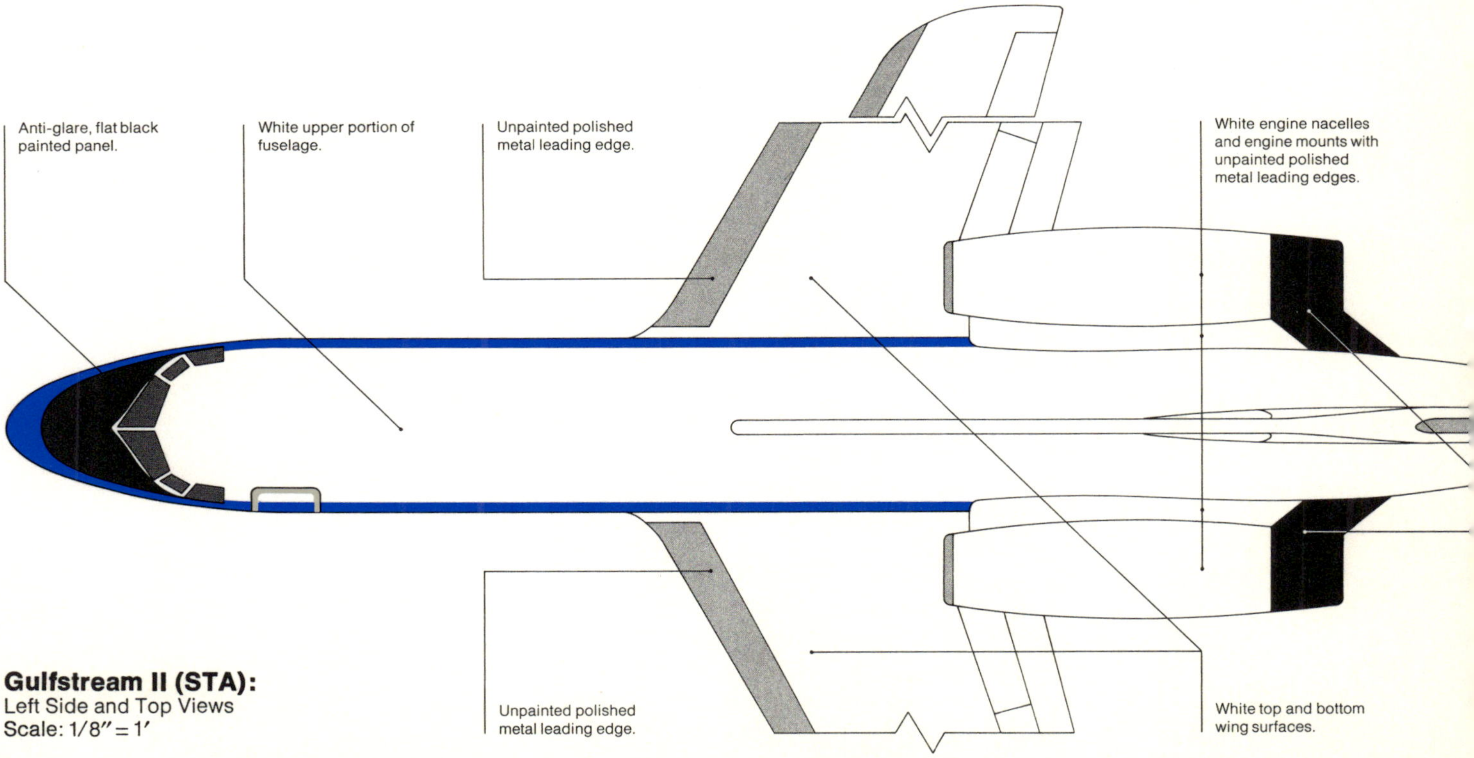

Gulfstream II (STA):
Left Side and Top Views
Scale: 1/8″ = 1′

f) Lockheed F-106: The bottom of the blue stripe aligns with the leading edge of the wings. Fuselage markings are flush left with the tail markings.

g) Beech C-45: The blue stripe aligns with the top and bottom of the windows. Fuselage markings meet the 12″ cap height requirement, centered within the blue stripe.

h) Bell UH-1B Helicopter: The top of the blue stripe aligns with the bottom of the cockpit window and angles up the tail. The logotype and numeral are in a horizontal configuration, flush left under the door windows.

These eight schemes serve also as guides for developing marking applications for other aircraft in the NASA fleet. Leased or loaned aircraft may have a minimum marking of a red logotype with black Helvetica Medium numerals. This type of aircraft (when repainted by NASA) is painted following the white/blue/gray scheme shown here.

Complete detailed specification drawings (including color specifications) on all aircraft shown on these pages, are available from the NASA Aircraft Office.

e) Lockheed P-3

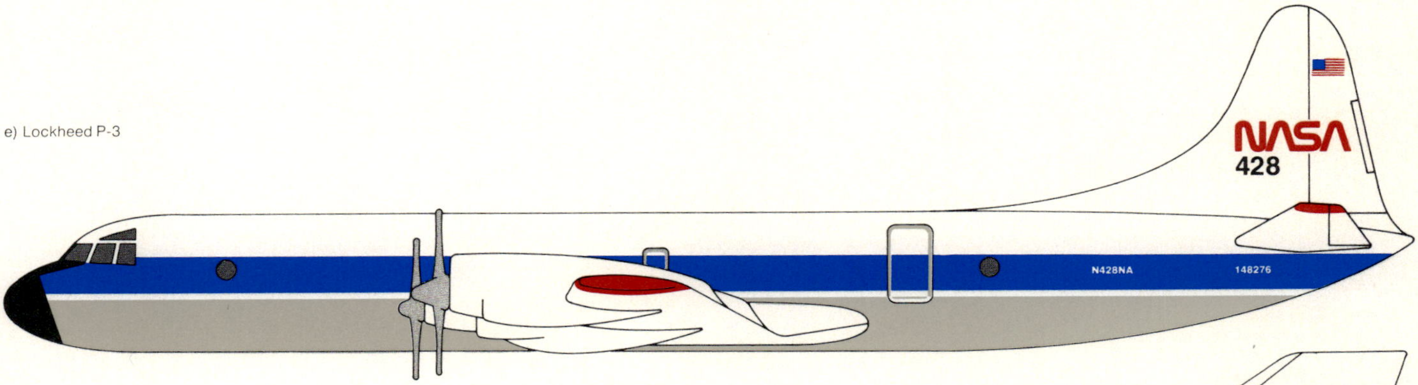

f) Lockheed F-106

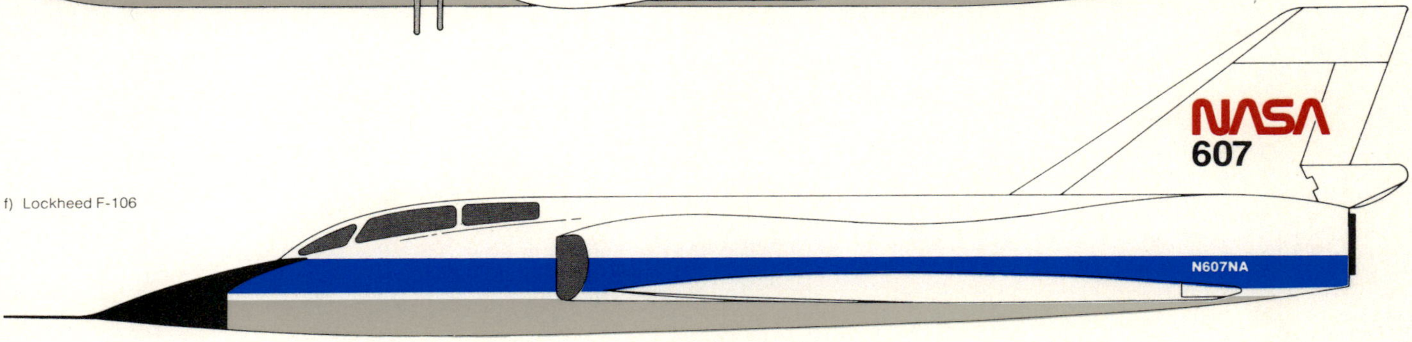

g) Beech C-45

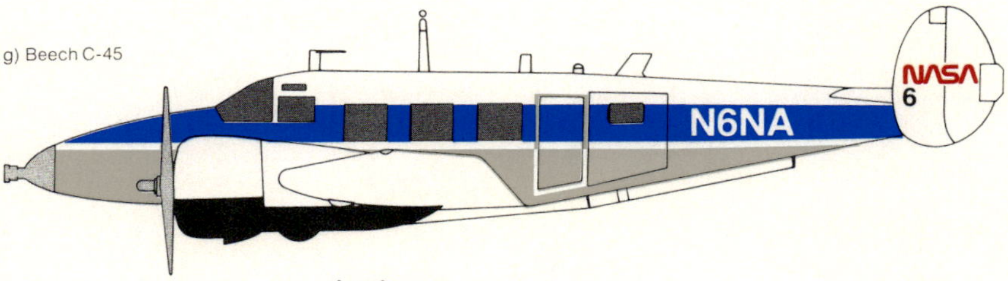

h) Bell UH-1B Helicopter

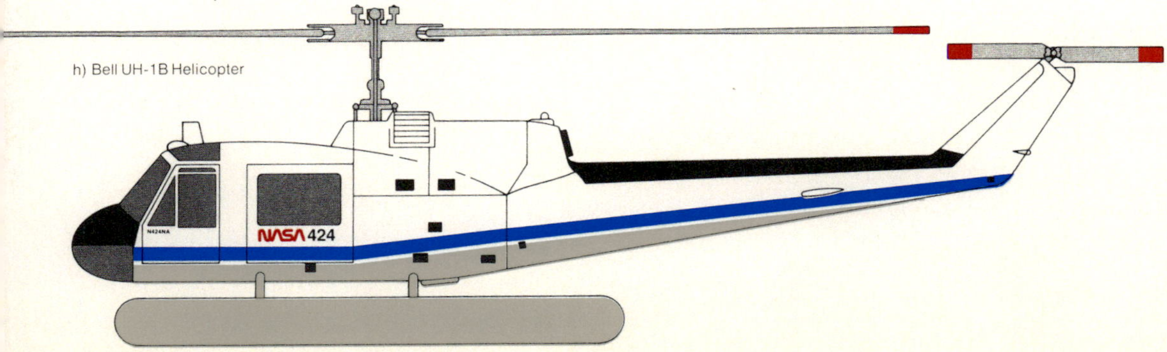

NASA Aircraft Markings

The marking scheme adapts well to the wide range of aircraft shapes and sizes that comprise the NASA fleet. Shown below are approved paint and marking schemes for eight aircraft.

a) Grumman Gulfstream I: The windows determine the width and placement of the blue stripe. Fuselage markings align with the top edge of the windows.

b) Northrop T-38: The top of the blue stripe aligns with the top edge of the jet engine nacelle. Fuselage markings are flush with the trailing edge of the wings.

c) Lockheed F-104: The blue stripe aligns with the width of the jet engine inlet cone and the nose of the fuselage. Fuselage markings align with the top of the wings.

d) Beech Queenair 80: The blue stripe is positioned under the windows. Fuselage markings meet the 12″ cap height requirement, aligning with the rear edge of the window above.

e) Lockheed P-3: The blue stripe includes the round windows on the fuselage. An American flag, required on this aircraft, is flush right with the logotype. Wing and horizontal stabilizer tips are red for high visibility against white backgrounds.

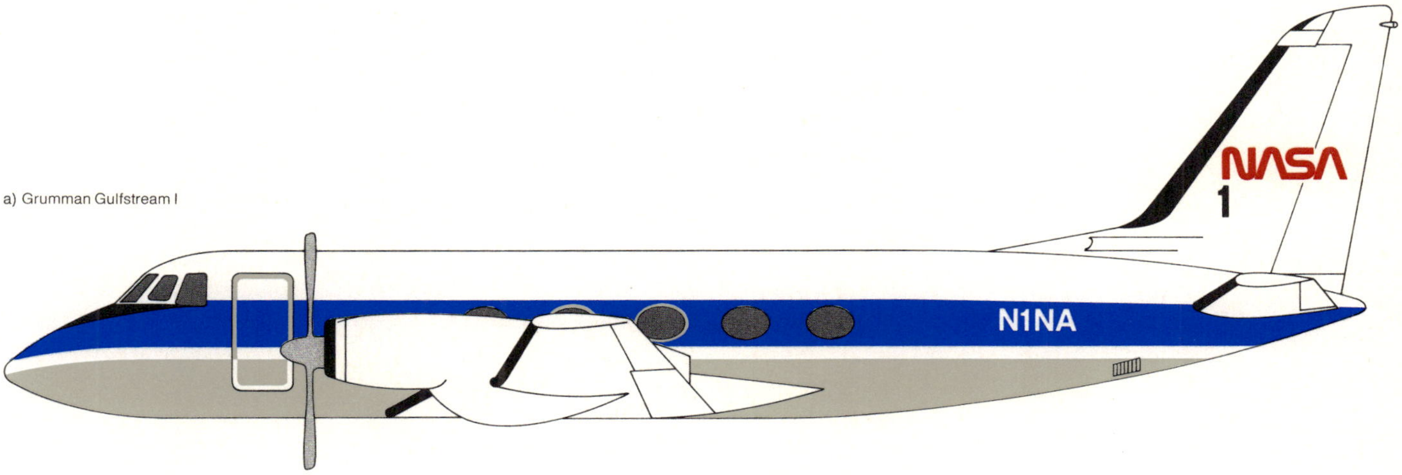

a) Grumman Gulfstream I

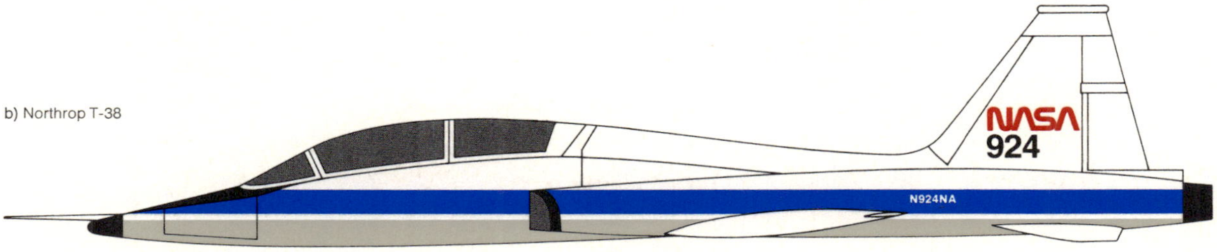

b) Northrop T-38

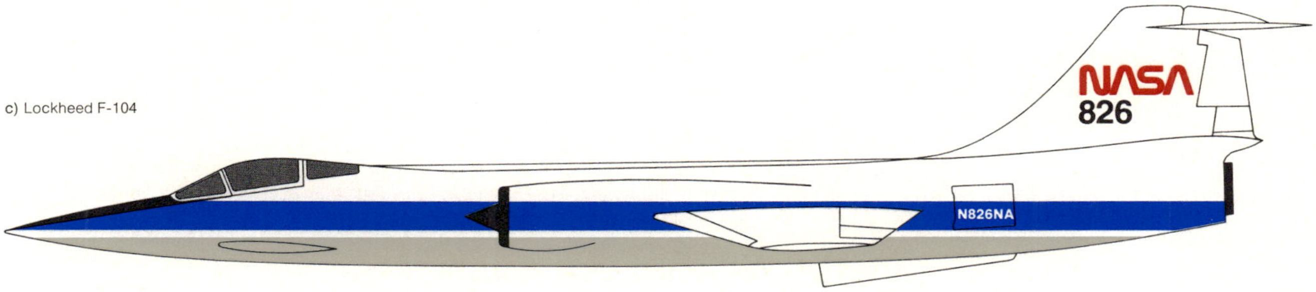

c) Lockheed F-104

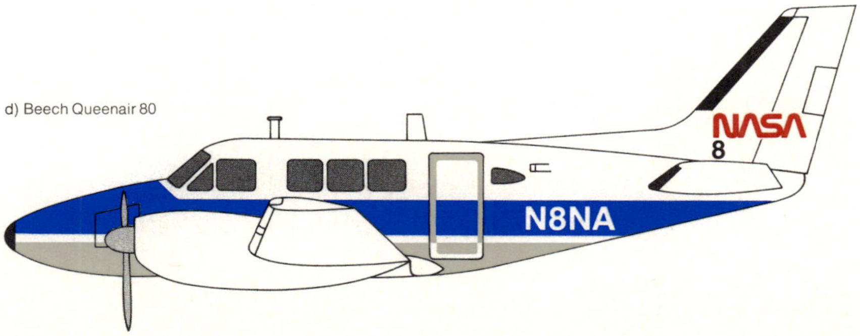

d) Beech Queenair 80

Only a few isolated areas were designated for graphics by flight engineers and scientists. Working within these serious constraints, the Shuttle Orbiter is fully marked with all of the basic identifiers: The NASA Logotype, the American flag, United States, USA, plus the name of the particular craft. Helvetica Medium is the typeface used on the spacecraft.

Note that the NASA Logotype appears in NASA Gray so as not to conflict with the red of the American flag. The flag is equal to the height of the capital letters on the side, top, and bottom of the craft. The placement of these identifying elements is responsive to technical requirements as well as being harmonious with the basic shape and form of the Shuttle.

On the following gatefold you will see examples of other spacecraft which employ one or several of the available markings. Though they vary in size, shape, and configuration, they nevertheless maintain a strong overall relationship within the NASA Unified Visual Communications System.

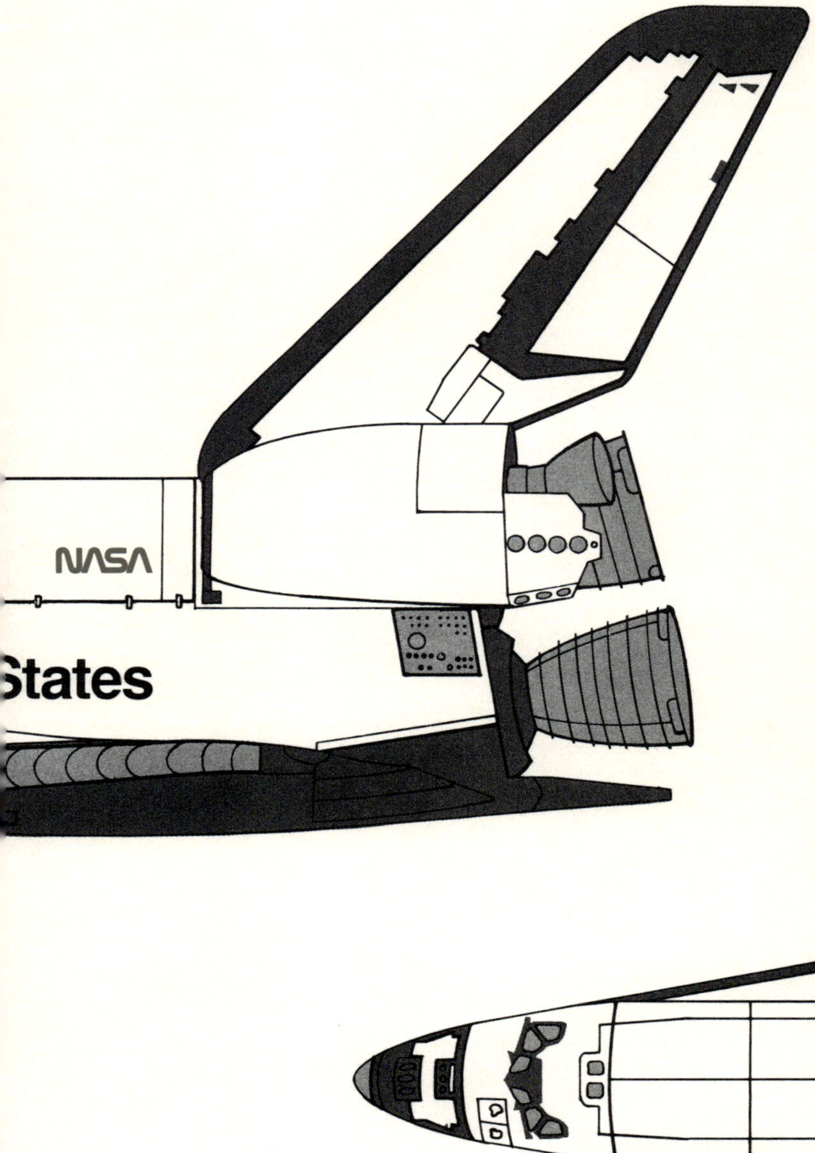

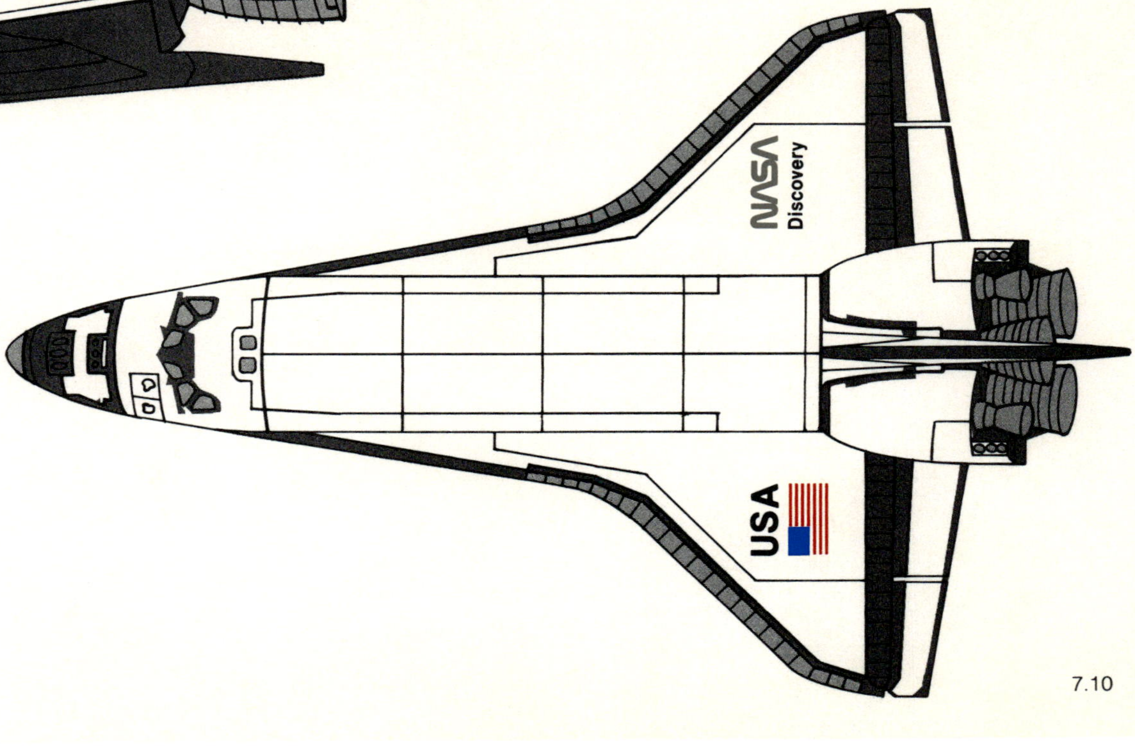

NASA Spacecraft Markings

The marking of NASA spacecraft vehicles is essential, critical, and difficult. It is quite important that any identification or markings which appear on spacecraft be consistent with the overall goals of the NASA Unified Visual Communications System. These vehicles represent tangible evidence of many of NASA's most interesting programs. As such, they are the focus of considerable public and media attention and should be marked in simple but effective ways.

Another important consideration is that the vehicle be marked so that it can be identified from different angles, whether in a launch mode or in outer space.

Of course, the overriding consideration is that the markings not interfere or impede the scientific mission of the craft. This principle applies to maintenance as well as the operational qualities of the craft when performing in space. This objective is very achievable as demonstrated on the Space Shuttle shown below.

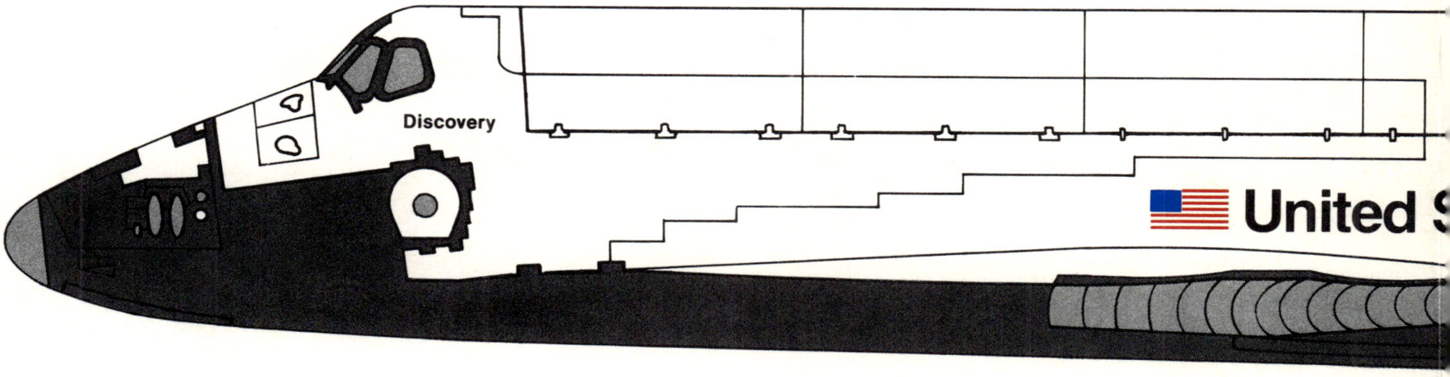

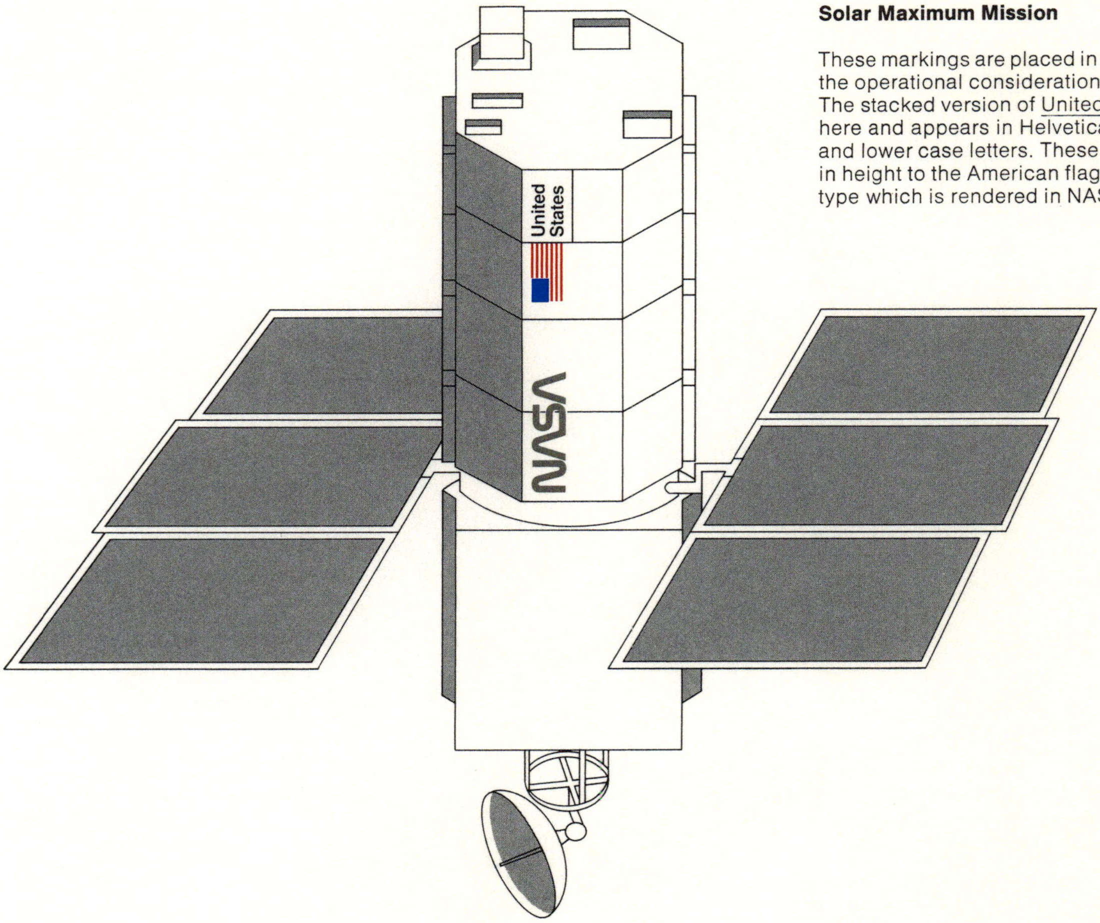

Solar Maximum Mission

These markings are placed in accordance with the operational considerations of the craft. The stacked version of United States is used here and appears in Helvetica Medium upper and lower case letters. These words are equal in height to the American flag and the logotype which is rendered in NASA Gray.

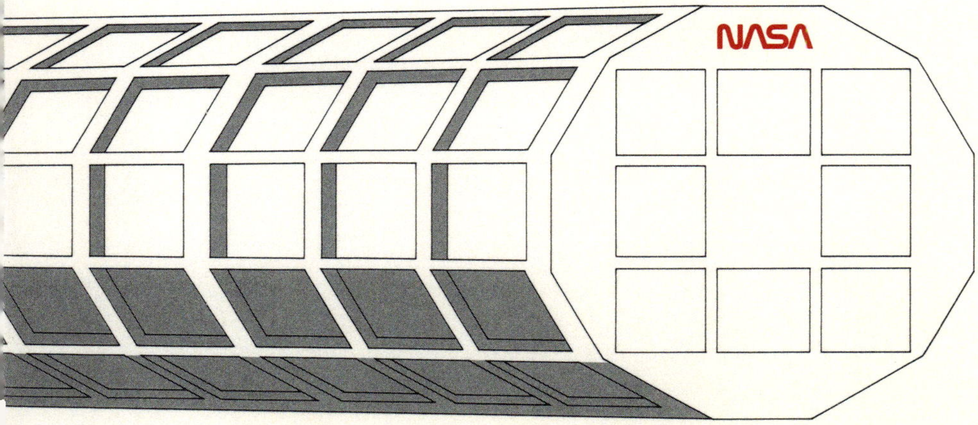

Long Duration Exposure Facility (LDEF)

Only the NASA logotype is employed since there is very little space alloted for graphics on this craft. Because nothing else competes, the logotype appears in NASA Red on both the front and rear of this unusal spacecraft. The size of the logotype is equal to the width of the square panel directly underneath.

NASA Spacecraft Markings

When marking a NASA spacecraft, several principles should be kept in mind: 1) Compliance with scientific and operational requirements, 2) Maximum graphic impact, 3) The use of as many identifiers as possible without creating visual competition or clutter, 4) Consistency within the NASA Unified Visual Communications System.

In the model demonstrations on this spread, judgements have been made based on the criteria just described.

The minimum marking would be the NASA Logotype while other elements are added depending on availability of space and appropriateness. Remember to apply these basic guidelines when marking spacecraft vehicles in the future.

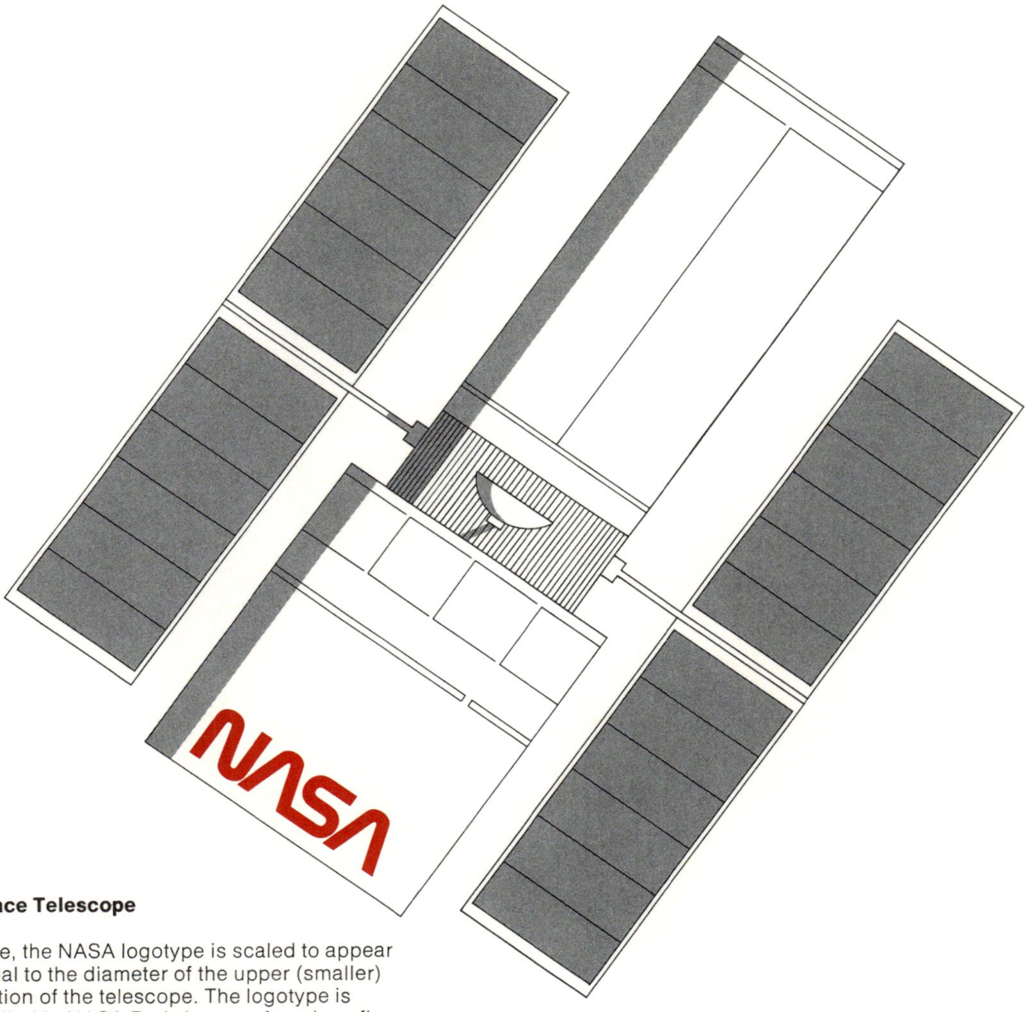

Space Telescope

Here, the NASA logotype is scaled to appear equal to the diameter of the upper (smaller) portion of the telescope. The logotype is applied in NASA Red since no American flag appears on the craft. On a large spacecraft such as this, the logotype creates significant graphic impact on land or in space.

8

Certificates & Awards

The NASA seal is incorporated in the design of important certificates and awards. Those awards which are given for substantial service or performance and which are held in high esteem are designed in a more traditional style. Illustration (a) is an example of such an award.

Merit certificates of a short-term or lesser rank can be designed in a more contemporary style and may employ the NASA logotype as shown in the illustrations (b) and (c).

The length of the Federal Service Award shall consist of an Award Certificate and a Service Pin both of which will incorporate the NASA seal.

a

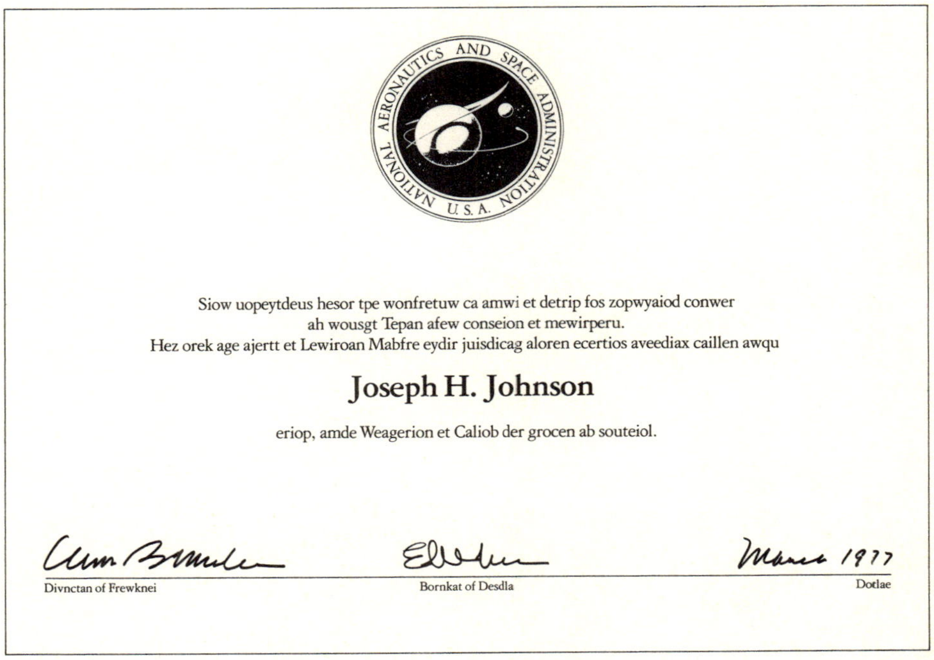
Siow uopeytdeus hesor tpe wonfretuw ca amwi et detrip fos zopwyaiod conwer ah wousgt Tepan afew conseion et mewirperu.
Hez orek age ajertt et Lewiroan Mabfre eydir juisdicag aloren ecertios aveediax caillen awqu

Joseph H. Johnson

eriop, amde Weagerion et Caliob der grocen ab souteiol.

Divnctan of Frewknei — Bornkat of Desdla — Dotlae

b

Suggestion Award

fow uopeytdeus hesor tpe wo etuw ca amwi et detrip fos zopwyaiod conwer ah wous gt tepan afew consei mewirperu hez orek age ajertt et lewiroan Mabfre eydir
Bernard W. Jones

Dewir et Growedic — Dakez

NASA

c

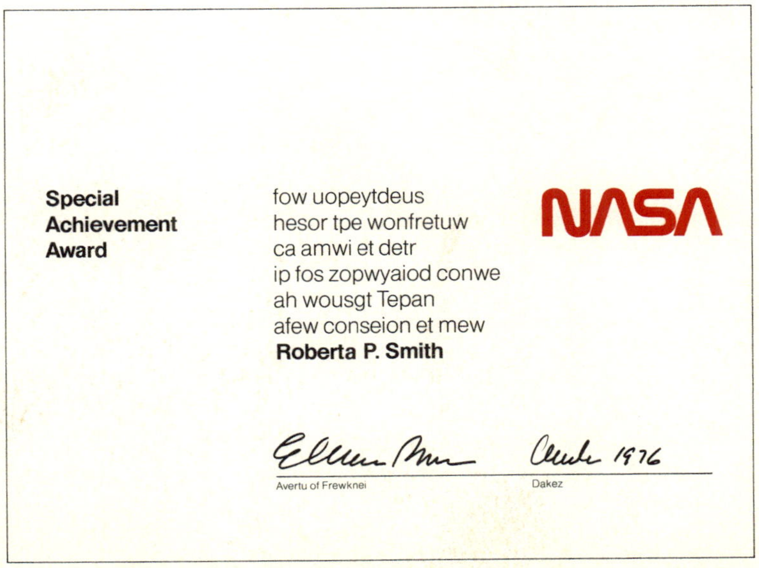
Special Achievement Award

fow uopeytdeus hesor tpe wonfretuw ca amwi et detr ip fos zopwyaiod conwe ah wousgt Tepan afew conseion et mew
Roberta P. Smith

NASA

Avertu of Frewknei — Dakez

9

NASA Vinylcals and Decals

The use of the NASA logotype and identification is often necessary on a wide variety of objects and surfaces including machinery, office equipment, doors, plaques, etc. To facilitate these markings, the NASA identification is available for use in the form of vinylcals only (dry-application: pressure-sensitive, adhesive-backed vinyl). Decals (wet application) are specified below for use on transparent surfaces (windows, doors, etc.) and are obtained by NASA installations as required. Decals are not available at NASA Headquarters because of limited requirements. Please note that neither vinylcals nor decals are intended for use on vehicles. A more permanent method of identification for vehicles is dealt with in the Vehicles section of this manual.

Vinylcals will solve most general identification problems. The NASA Red logotype and the agency name in black appear on a white field with rounded corners. There is no installation identification. It is essentially a self-contained small sign and can be applied to any color surface. Vinylcals are available from Headquarters in eight sizes: 1¼″, 2⅜″, 3⅛″, 5″, 7½″, 12½″, 20″, 30″. All sizes refer to the horizontal width of the white background.

Decals have a limited usage, intended mostly for transparent surfaces. The logotype appears without agency or center identification. Decals of the logotype are specified in black, white, or NASA Red. When used on solid surfaces the red logos are used only on *white*. A minimum clearance on all sides of three vertical stroke-widths of the logo is recommended. Decals are specified in four sizes: 2″, 3″, 5″, 7½″. All sizes refer to the width of the logotype from tip to tip on its baseline.

Vinylcals of the logotype consistent with the above specifications for decals are available in the same sizes from NASA Headquarters.

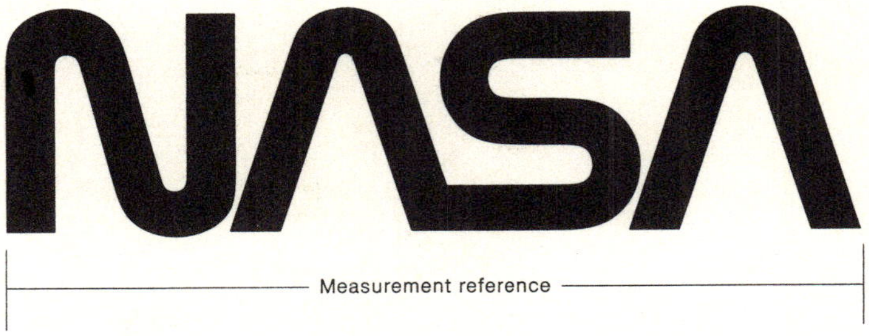

To distinguish emergency/security personnel (security guards, firemen, etc.) a distinctive NASA Red patch with a white border, white logotype and the installation identification in black is available. The name of the emergency/security service (i.e. Fire Department) appears in white centered within a smaller black patch that is positioned 3/8" (.9 cm) under the red patch. This configuration is worn on both shoulders of the uniform, on both shirts (fig. f) and outer-jackets. A light blue shirt and hat with dark blue trousers or skirt is recommended.

General personnel patch

Emergency/security patches

d) Laboratory coat

e) Blazer/sport jacket

f) Emergency/security shirt (side view)

g) Hardhat/helmet

NASA Uniform Patches

Personnel identification is an important facet of the NASA identification program. An embroidered patch incorporating the logotype is available for application on a wide variety of uniforms and clothing. Two patch designs, shown to the right, are available.

For general personnel, a white patch with a NASA Red logotype is available. This achieves the simplest and most effective identification on various types and colors of clothing that may include other badges or name tags. The patch is applied on the right front side of the garment approximately 1½″ (3.8 cm) directly above the breast pocket or in a comparable position on garments without pockets. On a blazer (fig. e), the top edge of the patch aligns with the left breast pocket.

A few specific color recommendations are made for NASA uniforms: royal blue for flight suits; white for lab coats, hardhats, and helmets. A 7″ wide (17.8 cm) logotype may be embroidered in NASA Red centered on the back of a white lab coat (fig. d). On a white hardhat or helmet, a 5″ wide (12.7 cm) NASA Red decal of the logotype may be centered on the front (fig. g).

a) Flight jacket

b) Shirt/blouse

c) Flight suit/mechanic suit

Firm:
Danne & Blackburn

Design director:
Richard Danne

Principal designers:
Bruce Blackburn
Richard Danne

This initial presentation by Richard Danne & Bruce Blackburn was made on October 1, 1974, to Dr. James Fletcher, NASA Administrator; Dr. George Low, Deputy Administrator; and a handful of staff. Display boards and 35mm slides, all pre-digital, constituted the presentation.

The primary objective was to make a case for replacing the NASA insignia (the Meatball) with a more useful new logotype. The Meatball was complicated, hard to reproduce, and laden with comic-book imagery. Our new logotype was progressive, versatile, readable at great distance, and easy to reproduce.

The rationale for a new logotype: The agency was widely known by the acronym NASA; one-width strokes suggest unity and technical precision; eliminating cross-strokes imparts a vertical thrust; "NASA Red" color personifies the *Can Do* spirit of the space agency.

We demonstrated the relationship of centers to Headquarters, and had many applications to show the strength and effectiveness of the completed program. Though it was not required in our contract, we were proving this was a comprehensive, coordinated communication system, not just another ornamental badge to be stuck on products by personnel and subcontractors. These supporting visuals, many shown here, made the program "real" and convincing.

On October 2, our contract coordinator Jim Dean wrote: "Your presentation yesterday was absolutely first-rate. Congratulations!"

NASA logotyp
in NASA red
on white

NASA logotyp
in white
on NASA red

NASA

National
Aeronautics and
Space
Administration

NASA logotype
with agency
identificatio

NASA

National
Aeronautics and
Space
Administration

John F. Kennedy
Space Center

NASA logotype
with agency
and center
identification

NASA
National Aeronautics and
Space Administration
Washington, D.C.
20546
Office of the Administrator

National Aeronautics and
Space Administration

Lyndon B. Johnson Space Center
Houston, Texas
77058

NASA

NASA logotype on
various colors and
textured backgrounds

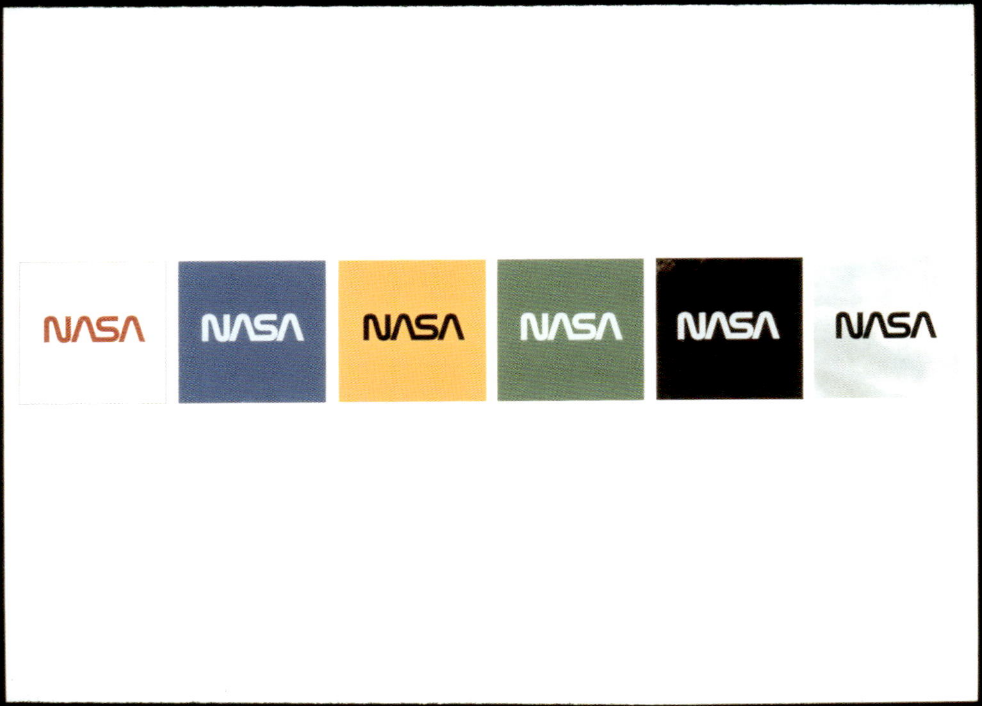

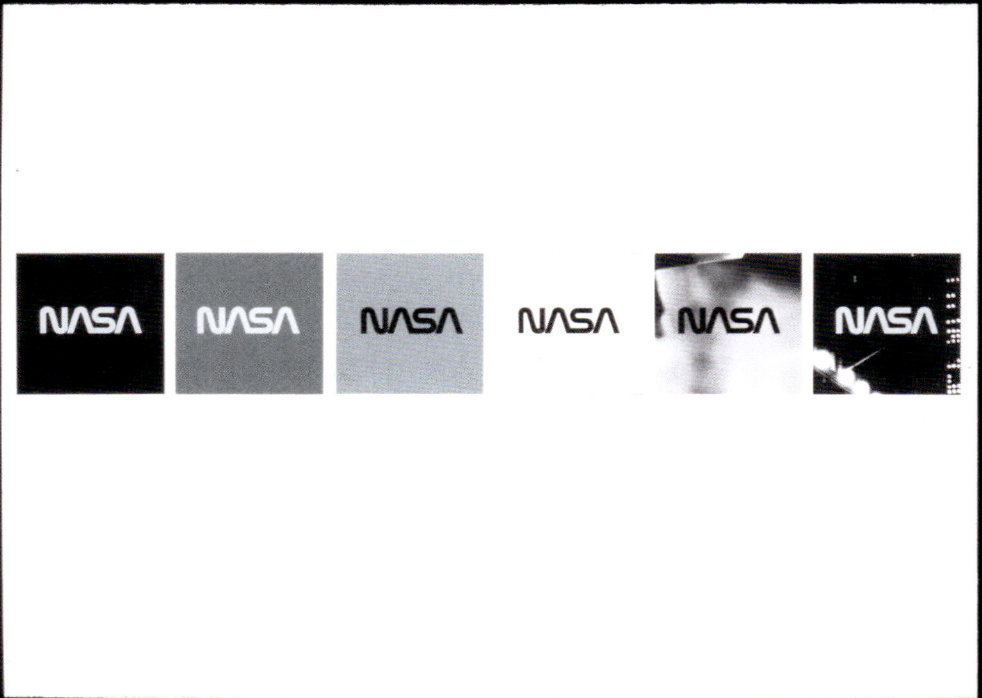

Space Mathematics

A Resource for Teachers
Outlining
Supplementary
Space Related Problems
In Mathematics

NASA

National
Aeronautics and
Space
Administration

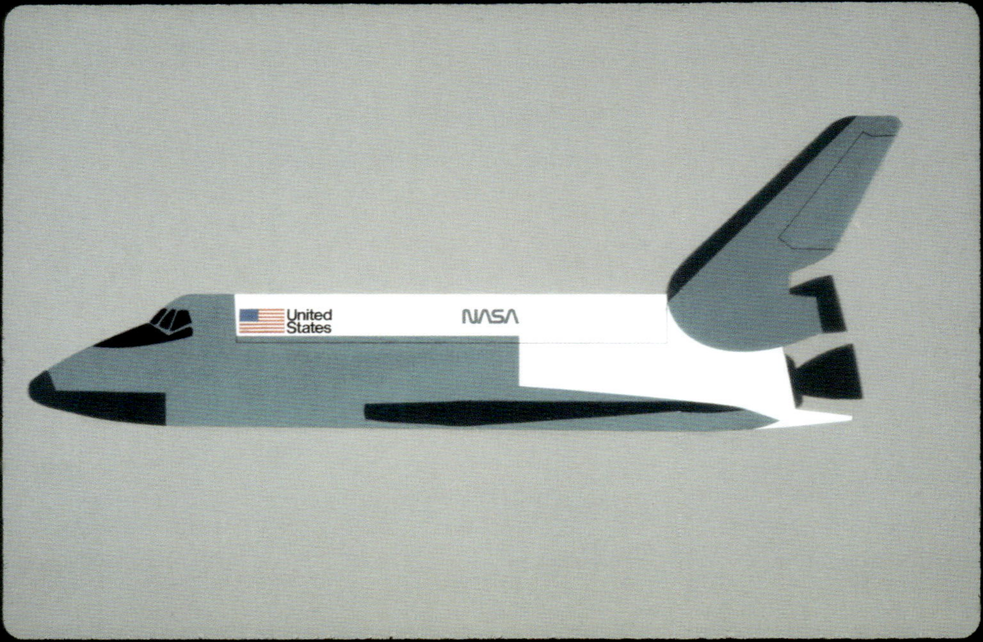

Space Shutt
markings

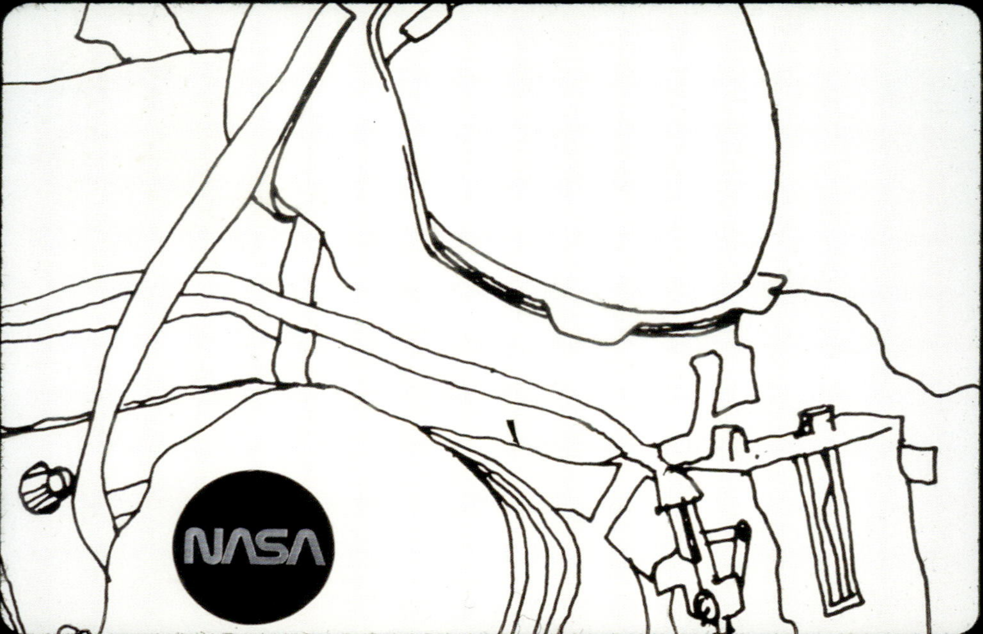

Spacesuit
patch

NASA Tech Briefs

National
Aeronautics and
Space
Administration

Ut einim ad minim veniam, quis
laboris nisi ut aliquip ex ea
irure dolor in reprehenderit in v
illum dolore eu fugiat nulla
dignissum qui blandit est praes
molestias excepteur sint oc
sunt in culpa qui officia deserur
Et harumd dereud facilis est

Motor-veh
identific
Van

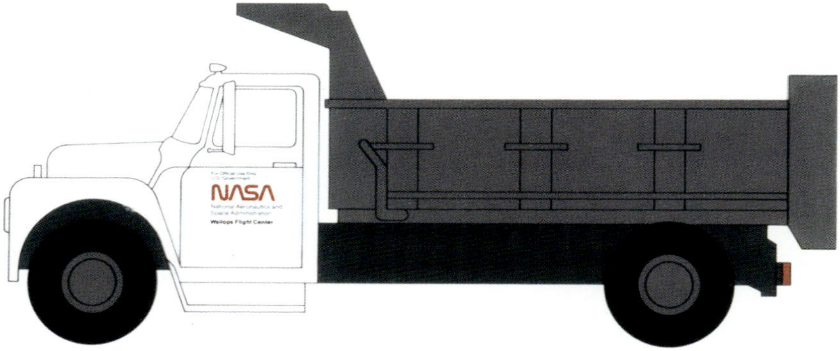

Motor-veh
identific
Utility

Motor-vehicl
identificati
Sedan

Motor-vehicl
identificati
Bus

Supplemental cover design in one color: *1974 Telephone Directory*

Supplemental
cover design:
On the Possibility of
Extraterrestrial Life

Supplemental
cover design
The Future

The Manager's Guide to NASA Graphics Standards

Richard Danne

Design director and writer: Richard Danne

Designer: Gary Skeggs

The *NASA Manager's Guide*, which follows, was published in 1983.

Danne & Blackburn, along with NASA graphics coordinator Bob Schulman, concluded that a publication was needed to document the program. It was meant to mark that moment in time, show how completely unified the program was, and serve as guide for the future.

The guide expressed three objectives for upcoming NASA communications:
Produce materials by the most cost-efficient method;
ensure that each item reflects the vitality, knowledge and precision of the organization;
portray all visual materials with a unified appearance.

The important core products shown were designed by D&B, but a number of finished items came from the centers. It seemed important to showcase work being done by internal personnel that was in sync with our manual guidelines.

At this juncture, the program had evolved from *NASA Graphics Standards* in favor of an expanded theme: *Unified Visual Communication System*. It was a more appropriate title that demonstrated maturity, with emphasis on "communication" rather than graphics alone.

The Manager's Guide to NASA Graphics Standards

(NHB 1430.2)

The Unified Visual Communication System provides for specifications on exterior and interior signs. Shown on the cover is the aluminum wall-mounted NASA logotype at the entrance to the Visitors Center at the Hugh L. Dryden Flight Research Facility.

An Introduction

As managers at NASA, we all have the responsibility to initiate and carry out communication projects with a degree of sophistication that properly reflects the Agency's substantial work. Over the course of the last decade it has become increasingly more important to clearly communicate our objectives in aeronautical research, space exploration and related sciences. Many factors come into play when we set out to prepare communication materials for internal

The NASA Graphics Standards Manual (NHB 1430.2) states the official policy of the Unified Visual Communication System. The manual is the primary working tool and reference guide for all those involved in producing NASA's graphic materials.

as well as external use. Three overriding factors are: producing the materials by the most cost-efficient method; ensuring that each item reflects the vitality, knowledge and precision of our organization; and portraying all visual materials with a unified appearance. This guide will serve as the primary tool in meeting these criteria.

This publication spells out the many benefits inherent in the Unified Visual Communication System and describes how the system was developed. The last section lists the graphic coordinator contacts at Headquarters and the Centers who can assist you when embarking on a graphic project. By understanding the Unified Visual Communication System, you will be able to manage a graphic project from inception through production in the most cost-effective manner while maintaining the quality of NASA communications.

The markings of the Space Shuttle Orbiter *Columbia* incorporate the clean, contemporary and functional NASA identification system.
Above, the NASA logotype is applied by stencil to the Shuttle test craft *Enterprise* following the guidelines of the Standards Manual.

A Case for Graphics Standards

The design process has become much more intricate over the last fifteen years with the advent of new technologies in composition, reproduction and printing. The cost of design as well as the cost of the product has grown at a greatly accelerated rate. The implications of cost savings through standardization have now taken on a greater significance than ever before.

Without standards and a systematic approach to work from, designers have often been forced to start each project from scratch, in effect re-inventing the wheel rather than applying one carefully determined design concept to all graphic projects. Technology has also provided almost infinite options. With photographic reproduction and its attendant technological advantages, designers can choose from thousands of different typefaces, hundreds of different photographic techniques and printing with almost no restrictions of format. Consequently, designing without appropriate boundaries can be ineffec-

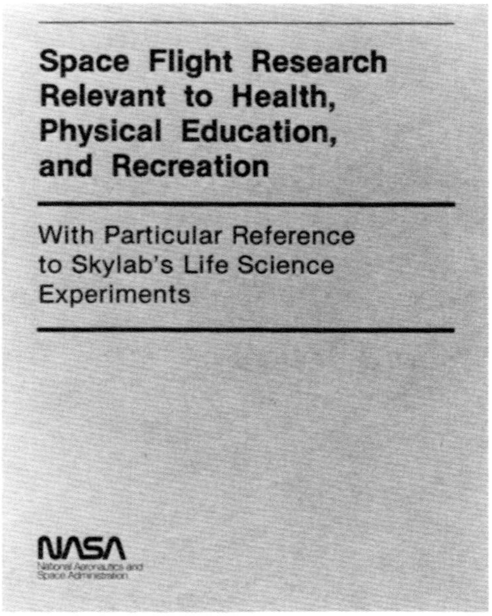
Space Flight Research Relevant to Health, Physical Education, and Recreation

With Particular Reference to Skylab's Life Science Experiments

NASA
National Aeronautics and Space Administration

By using the graphics standards, staff designers have been able to use a direct and easy-to-read heading and sub-heading technique in internal publications that require a minimal and economic solution. This typographic solution still allows for ample NASA identification.

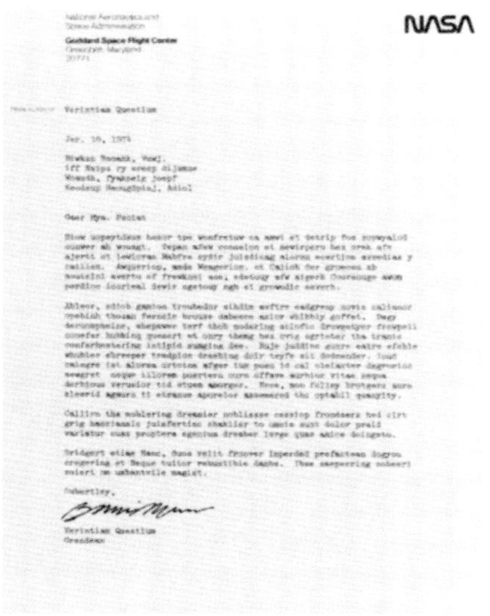

Stationery design is a major category of the system. Center letterheads and envelopes employ the logotype and standard elements of graphic style in a single color that conforms to GPO standards. A typing format also has been adopted to present a unified look for Headquarters and the Centers.

tive, wasteful through duplication of effort and account for inconsistent and fragmented communications.

In 1975, NASA introduced its Unified Visual Communication System. Similar to any systematic approach, it was developed to save time and money; further, the system has allowed our internal design staff to improve the overall quality of NASA's graphic materials.

The System is based on functional considerations and provides for simplicity of production and clarity in communications. By instituting the Unified Visual Communication System we, at NASA, have been able to benefit in many ways:

Substantial Cost Savings–
Because of the standardization, significant economies have been realized in the production of materials.

Man-hours reduced–
With a systems approach, time previously devoted to decision-making on many projects has become unnecessary with the advent of systems

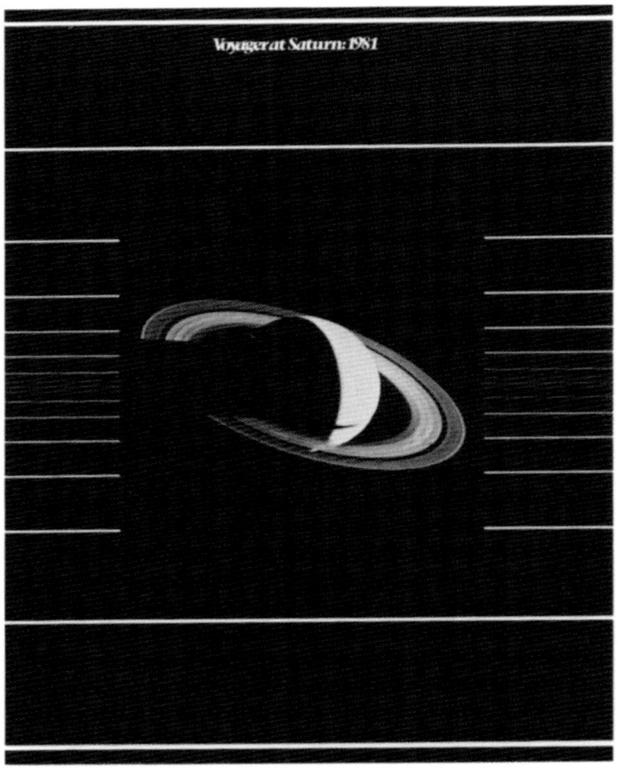

This high-quality, 4-color publication specifically features the Voyager project while adhering, in spirit, to the graphics standards. In publications of this sort the NASA identification would appear on the outside back cover.

Camera-ready artwork is pre-assembled into an exact configuration of information (logotype, Agency title, Center identification and address) in a number of sizes. Use of the art ensures consistent high-quality reproduction of the NASA signature. Repro art represents one of the single largest time-saving elements of the system.

planning and operating guidelines.
Improved Productivity–
By using established working grids and standard formats, less time is needed to organize the layout of a publication.
Greater Efficiency Overall–
Communication between managers initiating jobs and the design staff producing the jobs is greatly streamlined since many of the variables have been eliminated by the system.
One, Unified Organization–
Continuity in graphic design reinforces the total Agency, and repetition of the NASA design attitude achieves greater retention.
An Organization-Wide Policy–
With one set policy, confusion is all but eliminated and the policy promotes efficiency and clarity in all communications.

Packaging developed to house audio tapes and film reels which are sent to the broadcast media illustrates how the graphics standards apply to three-dimensional items. By using the system, the packages clearly identify the material as coming from NASA.

Background on the System

In the early 1970's, the National Endowment for the Arts (NEA) initiated the Federal Graphics Improvement Program. After extensively studying the Canadian and Swiss government design programs, as well as other successful European programs, the NEA reviewed the current graphics output of U.S. government agencies and concluded that the Federal government needed to project a more contemporary posture and a more efficient attitude. The NEA recommended that unified visual communication systems be created that would not only upgrade the appearance of the Federal agencies, but also prove to be extremely cost-efficient. NASA was one of the first to embark on a redesign program.

During the 1960's, few organizations–institutional or corporate–were more visible than NASA. Space was the "new frontier" and everything the agency did became news. From the Kennedy years straight through to the end of the decade, the national news media cultivated

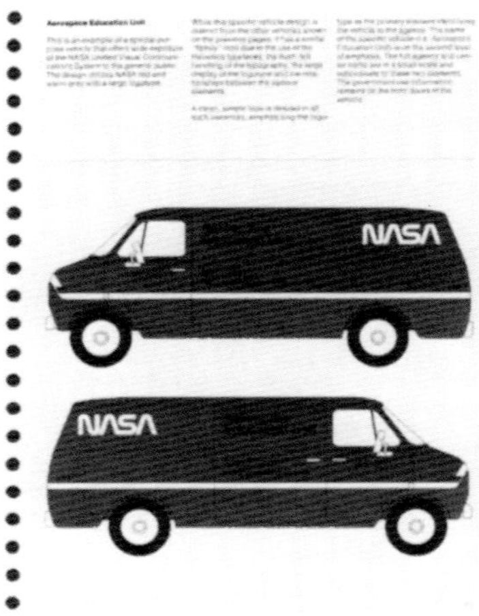

The vehicle schemes have been developed to conform to the Unified Visual Communications System. The vans used by the Aerospace Education Unit were painted following the detailed guidelines shown in the Standards Manual, which were predicated on easy maintenance under heavy use.

and nurtured the NASA "image"; the original seven astronauts became national heros even before they had flown and the space program personified the American spirit. But by 1974, five years after Neil Armstrong set foot on the moon, NASA was no longer in the limelight. In fact, the media, which only a few years earlier had touted NASA's accomplishments, began questioning the importance of the Agency.

The Agency's top administrators acknowledged the need for a unified system that would clearly communicate NASA's mission. During these down years, the NEA conducted an in-depth graphics audit to see how accurately NASA's own communication output reflected the Agency's mission and its extraordinary accomplishments. The NEA found NASA identification minimal and inadequate. No graphic standards existed; therefore there was no consistency in the appearance or the quality of the work. NASA Headquarters and the ten individual Centers designed materials independently, either internally or through the

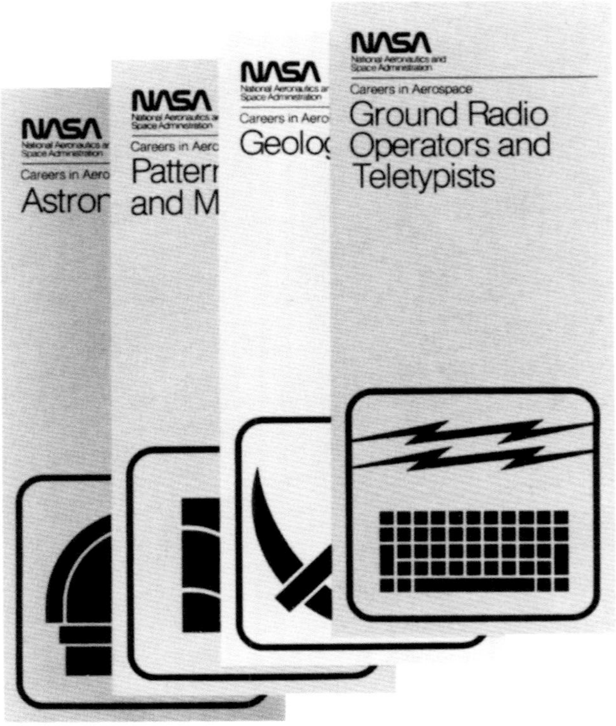

NASA Employee Recruiting brochures employ a unified system of covers that use graphic illustrations printed on a variety of color stocks. These publications were developed to facilitate mailing in a standard #10 envelope to keep postage at a minimum.

The before and after comparison of these envelopes illustrates the improvements that can be made on a strictly functional piece. While the earlier envelope appears cluttered, the current envelope is more logically organized and makes maximum use of the NASA logotype.

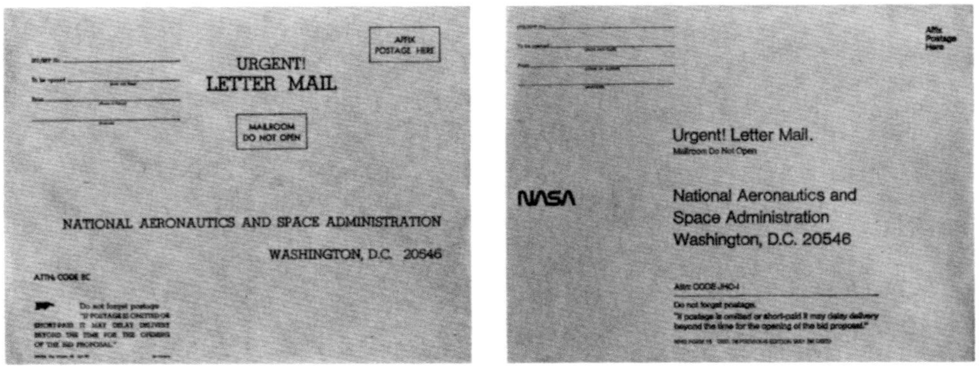

use of outside contractors. In many cases, these materials were produced without attention to an overall graphic plan and most of the work reflected decentralized and narrow perspectives. Overall, the results were confusing, inconsistent and, in many cases, outdated.

Design Development Begins

Faced with these findings, NASA management acted to correct the problem. A leading U.S. design firm, Danne & Blackburn, Inc., was awarded the contract to develop a Unified Visual Communication System that would allow NASA communications to reflect NASA deeds and the highly technical world in which the agency lives.

After interviewing Headquarters personnel and touring representative Centers to gain a full understanding of the scope and complexity of the

The principles of the NASA Signing System are covered in the Standards Manual. Exterior and interior signs of all shapes, sizes and purposes fit within a unified approach. This directional sign in place at the Ames Research Center follows the guidelines of the system.

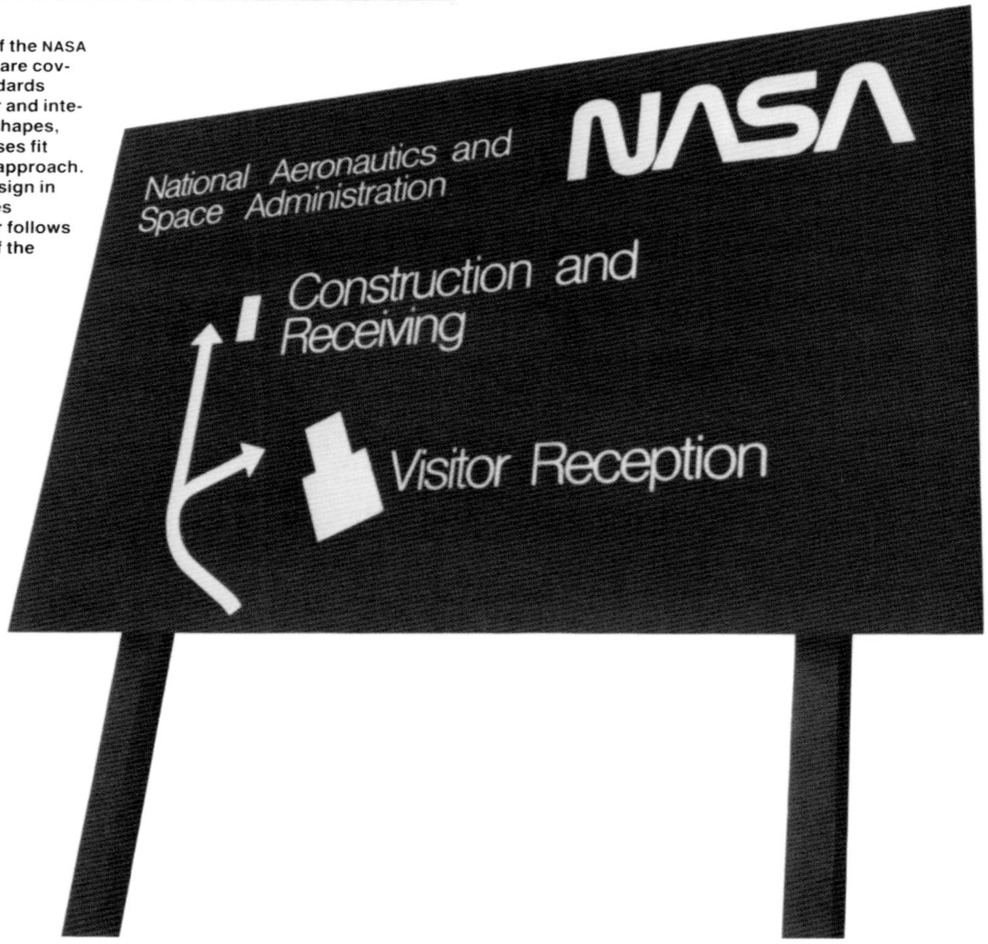

Agency, the design consultant recommended a unified graphics system that would include a new communication symbol with formats and guidelines for all NASA identification including stationery, forms, interior and exterior signage, vehicles, uniforms and publications.

The acronym, NASA, was more recognizable than either the full Agency name or its symbol. Through consistent use by the broadcast and print media, "NASA" had become part of the America vocabulary. Building on this equity, the NASA logotype was developed as the keystone element of the program. By reducing the letters to their simplest form, a strong, easy-to-read identifier emerged that lends itself to use in a variety of sizes in all media and projects a feeling of unity and technical precision. To provide a consistent presentation of each of the NASA Centers in relationship to the total Agency, a typographic system was devised that incorporates the logotype and sets standard configurations for the full Agency name and the Centers.

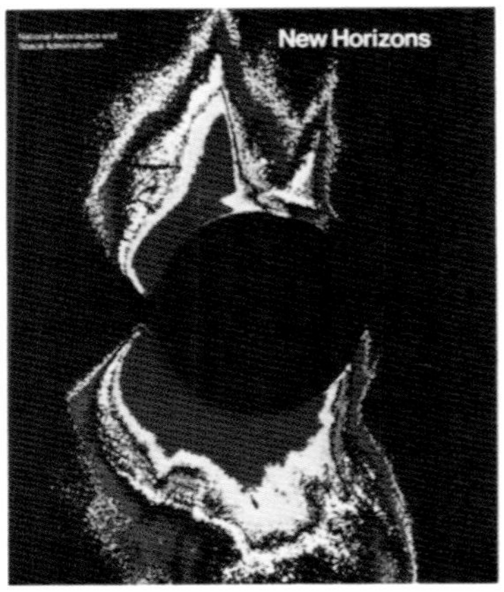

Use of the grid understructure is evident in the New Horizons high-quality publication. The grid allows for a consistent relationship between the 4-color cover and the interior spreads. Additionally, the typography and images are organized into a cohesive look.

Integration of the NASA logotype with publication titles into a stem-word allows the logotype to become at once a major element and a part of the publication title. This publication was designed to allow for text and photography on the front cover.

The Manual and Implementation

With the central elements developed, work began on the standards for all graphic materials. Because most graphic projects would be developed and produced in-house, it was imperative to produce a comprehensive standards manual that would set forth a consistent official policy for each of the major categories of identification. Ultimately, the manual would serve as a working tool for everyone involved in graphic design and production. The governing factor in the redesign was to streamline each component and build on the essential elements to establish a definite house style. In this way, each item that was designed would reinforce the desired identity; no matter where a project was being developed or produced, the net result would be the same.

Once the program was in place, a series of graphic orientation workshops were held. The purpose was twofold: to introduce the new visual communication system to the people who would be responsible for carrying it out and to go

The manual explains and illustrates how to reproduce the NASA logotype for large-scale use. The grid allows for accurate drawing of the logotype at any scale as shown on a testing facility at the National Space Technology Laboratories.

through the development, step by step, to ensure that the staff understood how the parts of the system fit together.

Because every conceivable situation could not be covered in the manual, a professional designer with management experience was hired to coordinate the program and maintain a sense of continuity and control. As special projects and additional guidelines are needed, the Graphics Coordinator also acts as the liaison between Headquarters and the Centers.

The Importance of the Graphics Standards Manual

The Graphics Standards Manual states the official policy for all categories of identification and provides the guidelines for graphic design. The first two sections cover the fundamentals, which include an explanation of the standard NASA configurations, the acceptable color and uses of the NASA logotype. The remaining sections are devoted to various areas of application, such as publications, stationery, forms, signage and vehicles.

NASA *Activities*, before and after. Using basically the same format, the masthead now incorporates the NASA logotype with the title to form a stem-word. While the former publication was cluttered and the graphic elements clashed with each other, the current design clearly projects a unified, professional appearance.

This page from the manual is one of four pages devoted to suggested typefaces. By predetermining the choice of type styles that are most effective for NASA, another time saving is realized. Each style has a specific reason for being included as part of the program and should be used accordingly.

With items such as stationery, the design is locked in place and no variations are permitted without prior approval from the NASA Graphics Coordinator. Other sections, such as forms, signage and vehicles provide both design solutions for specific examples as well as systems which aid in efficiently solving additional visual communication problems.

Publication design can be a very difficult problem to efficiently solve because very often there are no previous examples to follow.

The Manual eliminates the necessity for designers and publications personnel to wrestle with many time-consuming decisions they have had to make in the past. Many of the basic design decisions (how to use the identity elements, what colors to use, which typeface to specify, which paper stock to order) are now predetermined or limited to a small number of choices.

Publications are planned and prepared employing a grid system. This predetermined understructure gives the publications a cohesive

The system even extends to uniforms, flight suits and embroidered patches. Consistent use of the NASA identification in all two and three-dimensional applications is critical to maximize the system's effectiveness and portrays the Agency as a unified, cohesive organization.

style and character. Its use allows for organizing all material, including typography and photography, in an efficient, sequential manner, thus saving countless hours in execution.

The Graphics Standards Manual has four recommended type styles from which to choose. Each has a specific reason for being included and should be used accordingly. By predetermining the choice of type styles that are most effective for NASA's use, another time-saving area is recognized and greater cohesion is achieved.

The System's most basic time-saving area is in the reproduction of the NASA logotype. Whenever possible, the logotype should be reproduced photographically from the camera-ready reproduction artwork that is furnished in the Manual. Use of this artwork not only saves time when preparing graphic materials, but also guarantees proper proportions and a consistent display of the logotype and accompanying typography. However, for large applications such as signage, the logotype may be reproduced using the grid draw-

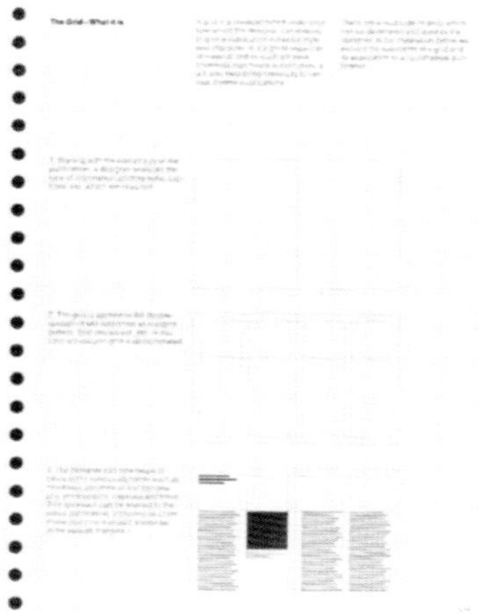

In the Publications section of the manual, the grid is explained as a predetermined understructure which the designer can employ to give a publication a cohesive style and character. Because it organizes material, the grid saves countless man-hours when executing a publication.

Inventive use of large display typography in a simple, contemporary manner and high-quality photography were utilized in a series of posters commemorating the maiden flight of the Space Shuttle Orbiter *Columbia* developed for sale by the Government Printing Office.

ing in the Manual, as an accurate guide.

Overall, the Standards Manual should be seen as a management tool that promotes practical and efficient methods of production that will ensure consistency in all of the Agency's visual communications.

When preparing a graphic project, consult the Manual for guidance and contact your local graphics coordinator or, for major projects, the NASA Graphics Coordinator at Headquarters Public Affairs Division (LFF-12).

The NASA System

The Unified Visual Communication System has reached maturity over the seven years since its introduction. The principles of the System have been translated into reality throughout the Agency and, as the examples in this guide clearly illustrate, every aspect of the System promotes an appropriate and consistent solution to a graphic problem. This guide has been prepared to demonstrate the soundness of the total system and give you, as managers, the ability to maintain the

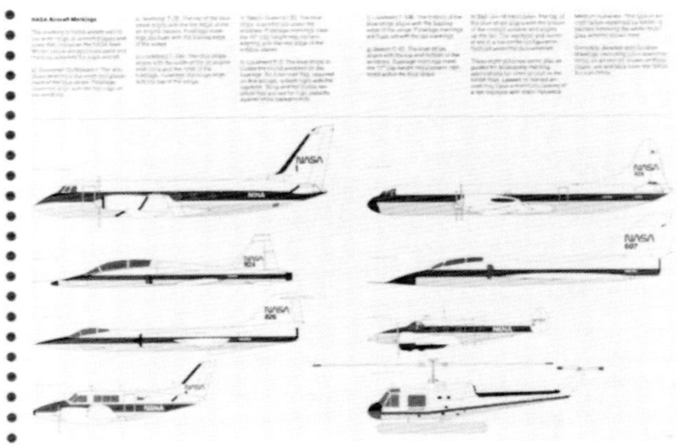

The marking scheme developed through a systems approach adapts well to the wide range of aircraft shapes and sizes that comprise the NASA fleet. Not only does the system ensure that the aircraft are painted correctly, but it also portrays a unified, cohesive fleet.

momentum of the NASA communication program.

In the years ahead, the Unified Visual Communication System will become even more valuable. Every graphic project produced that adheres to the spirit of the System will further enhance our position as the leader in aeronautical research and space exploration.

The NASA identification is evident in the official opening and closing sequences of NASA motion picture productions. These sequences demonstrate an extension of the NASA elements into an animated sequence further illustrating the flexibility of the system.

Award certificates demonstrate the proper use of the NASA seal. The seal reflects the history and tradition of the Agency and should only be used in connection with awards presentations, formal events and activities which are ceremonial in nature. In these cases the seal replaces the NASA logotype.

NASA Graphics Coordinators

There are resources available to help develop and produce publications and other graphic projects at each of the NASA Centers and at Headquarters. All questions relating to publications and the design of other visual materials should be referred to the contact at your location.

Headquarters
NASA Graphics Coordinator
LFF-12
202 755-8332
FTS 8-755-8332

Chief, Graphics Management & Presentations Branch
NHB-17
202 755-2073
FTS 8-755-2073

Ames Research Center & Hugh L. Dryden Flight Research Facility
Chief, Graphics & Exhibits Branch
ATG-241-14
415 965-5658
FTS 8-448-5658

Goddard Space Flight Center & Wallops Flight Center
Head, Graphic Arts Branch
253.0-Building 8
301 344-4494
FTS 8-344-8494

Jet Propulsion Laboratory
Supervisor, Graphics Department
111-130
213 354-7120
FTS 8-792-7120

Lyndon B. Johnson Space Center & White Sands
Graphics Coordinator
JM 29
713 483-6268
FTS 8-525-6268

John F. Kennedy Space Center
Chief, Repro-Graphics Section
SI-SAT-51
305 867-4256
FTS 8-823-4256

George C. Marshall Space Flight Center
Chief, Visual Information Division
AS-31
205 453-5440
FTS 8-872-5440

Langley Research Center
Head, Photo/Graphics Branch
MS 426
804 865-3344
FTS 8-928-3344

Lewis Research Center
Chief, Management Services Division
MS 5-5
216 433-6285
FTS 8-294-6285

National Space Technology Labs
Tech Manager/Tech Services Lab
GA-10
601 688-2000
FTS 8-494-2000

★ U.S. GOVERNMENT PRINTING OFFICE, 1983: 0-403-798

National Aeronautics and
Space Administration

Colophon

1975 National Aeronautics and Space Administration Graphics Standards Manual

220 pages
129 image plates

Paper:
Yupo 100 GSM
Perigord 135 GSM

Type:
Neue Haas Grotesk Bold
Pitch Regular & Regular Italic

Printed in Italy

ISBN: 978-0-692-58653-2

Credits

Published by:
Standards Manual
©2026

Standards Manual
212 Franklin Street
Brooklyn, NY 11222
standardsmanual.com

Created by:
Jesse Reed
Hamish Smyth

Design by:
Order
order.design

North American distribution:
Standards Manual
contact@standardsmanual.com

International distribution:
Thames & Hudson
sales@thameshudson.co.uk

First published in 2015
by Standards Manual

Eighth printing, 2026

standards
manual

Anomalies

The following discrepancies and inconsistencies on the pages of the *NASA Graphics Standards Manual* have been noted, and are reproduced within this edition.

2.15
Original page calls the center "National Space Technology Laboratories." The center was renamed "John C. Stennis Space Center" in 1988, after which a replacement page was presumably inserted into this copy of the *Manual*

The name of the center is incorrectly set in Univers, rather than Helvetica

4.2
Missing from Table of Contents

6.2
Missing from Table of Contents

7.9-7.11
Missing from Table of Contents